Preserved Railway Coaches

Michael Harris

LONDON

IAN ALLAN LTD

First published 1976

ISBN 0 7110 0664 4

Published by Ian Allan Ltd, Shepperton, Surrey,
and printed in the United Kingdom by
John G. Eccles, Inverness

*To my wife Sheenagh for all
her help and encouragement
in the preparation of this book.*

Title page: *NER Saloon No 1661 in use on the KWVR.* C. P. Boocock.
This page: *Restored GWR corridor stock on the SVR.*

Contents

Foreword

The purpose of this Foreword is to make sure that all those who have helped to make this book a comprehensive record — hopefully — of preserved passenger stock in Britain and Northern Ireland are given due recognition at the start rather than the finish. If vehicles have been omitted then it is my responsibility alone; so are any errors of fact. Unfortunately it has not been possible to find suitable photographs for all the coaches described — in time the owners may like to forward to me prints of restored vehicles in all their finery.

I would like to thank most warmly all those who have answered my enquiries so fully, and the many organisations who have provided many useful photographs. I am particularly indebted to the Eastern Region of British Railways, W. R. Devitt (Flying Scotsman Enterprises), John Hosegood (Great Western Society Ltd), Brian Kohring, S. W. Smith (Midland Railway Co Ltd), Barry Bull (Railway Preservation Society), Charles Friel (Railway Preservation Society of Ireland), Andrew Harper (Scottish Railway Preservation Society), Ron Black (Strathspey Railway Association) David Jenkinson (National Railway Museum, York) and R. Beggs (Ulster Folk & Transport Museum). A. B. Macleod was also most kind in helping me to locate some interesting photographs in the Ian Allan photographic library.

Introduction

In the stages of planning this book it seemed remarkable that it would eventually contain references to over 600 coaches, railcars, electric multiple-unit coaches and passenger stock vans. The British Transport Commission list of preserved passenger coaches of 1956 totalled some 30 vehicles. Since then, and more particularly since the early 1960s, the railway preservation movement has grown to an amazing extent so that the collection of passenger vehicles very closely reflects the variety of changes in rolling stock design from the 1830s to the late 1950s. One must also pay tribute to a relatively small number of people who believed that rolling stock should be preserved and, having acquired it, saw that it would be restored. A number of vehicles may be preserved over the next few years but it would seem that the rate of acquisition must decline. This is also a convenient moment to suggest that in looking at the large number of coaches awaiting restoration, the preservation societies might consider providing grants to the less prosperous organisations to avoid deterioration of some valuable vehicles.

Official preservation has been hampered by a shortage of money and an unhappy wrangle in the mid-1960s over the future of the national railway museum. Hopefully this is now past and it is encouraging that the British Railways Board, the Science Museum and the societies are working together — as for example with the loan of the GWR Buffet Car No 9631 to the Severn Valley Railway. A similarly good development has been the agreement of the British Railways Board that privately-owned vehicles could, subject to approval, run over British Railways lines — reference is made to this in the description of individual vehicles.

Special tribute should be paid to the preservation situation in Northern Ireland and Eire, where the Ulster Folk and Transport Museum (previously the Belfast Transport Museum), Coras Iompair Eireann, Northern Ireland Railways, Irish Railway Record Society and Railway Preservation Society of Ireland have all worked closely together in the preservation of vehicles and collation of historical material.

So this book is an attempt to show the diversity of coaches preserved, the enthusiasm and expertise of the preservation movement, and the fact that the collection of over 600 coaches is a priceless national asset which is part of our railway heritage.

The 600
Not all the coaches in this book have been acquired for the same reasons. Some of the preserved railway operators have purchased BR standard stock for their scheduled services rather than rely on the necessarily slow rate of restoration of older, more historic vehicles. These BR coaches have required minimal work to get them into traffic. Even so, many must be regarded as historic. Other companies such as the Severn Valley and Bluebell use 'historic' stock in everyday service. Such vehicles, with their wooden-framed bodies and steel panelling, may be easier to maintain in the long run because they are simpler to repair than all-steel stock. In 1974 BR introduced a system of approving privately-owned coaches for running over their lines on steam or diesel specials. These vehicles are then given plates with BR serial numbers to denote their prowess. A number of coaches in the book are therefore referred to as 'plated' — a new, select band of preserved vehicles.

Finally, a number of other coaches have been preserved as static exhibits as perhaps their interiors are unlikely to be reinstated except at very great expense. All these policies on preservation are relevant and they are worth stating so as to put the picture in perspective.

What we have Lost

The diligent detection of such a wide range of vehicles has meant that there are remarkably few gaps in the development of the British railway coach from the 1830s through to the present day. Most coaches preserved before the 1950s were exceptional examples such as royal saloons. In the meantime a number of rarities disappeared. It is sad that no complete GWR broad-gauge coach has been preserved — in view of the destruction of the broad-gauge locomotives *North Star* and *Lord of the Isles* by the GWR in the 1900s perhaps this was not surprising. We have no examples of 'parliamentary' coaches — possibly the railways were too ashamed to keep them!

One of the Metropolitan Railway's rigid eight-wheelers — an example survived on the Brill branch until 1935 — is another notable omission. A Great Northern/East Coast Joint Stock clerestory twelve-wheeler would have been a very worthwhile candidate for posterity; so would a Caledonian 'Grampian Corridor' coach, of which a few survived in Army ownership into recent days.

Otherwise there is little to complain about until a more recent era. The period 1960-5 saw the virtual elimination of coaches of the 1920-40 era. Unhappily, the streamliner stock of the LNER and LMS was allowed to go for scrap, apart from the two 'Coronation' observation cars. It was a pity that the BRB made no attempt to rescue examples such as one of the 'Coronation' twin open firsts or the unique 1939-40 'Coronation Scot' club car.

There have, of course, been one or two examples of wanton destruction. A London Tilbury & Southend coach was carefully restored for the Tilbury line centenary, included in the BTC's list of preserved coaches for 1956 and then scrapped. So was a Wisbech & Upwell tramway coach — although the body of another has since been located. In addition, an early tube-car from the Great Northern, Brompton & Piccadilly Railway — motor car No 51 of 1906 — was preserved by LT, but then scrapped although part of the body is in the Syon Park museum.

But, considering what might have been lost, we must be very grateful.

Preservation in the Future

It is only a suggestion, but it would appear to make good sense to draw up a list of vehicles still in service which should be reserved for preservation in the future. This might be done by co-operation between the Science Museum — now responsible for the national railway collection and the National Railway Museum at York — and the railway preservation societies. The following is offered only as a personal list of preferences but seems to include some of the more notable candidates:

East Coast Joint Stock Royal Saloons Nos 395/6 1908
LT 1938 Stock Tube car (already proposed by LTE)
Wirral/Southport electric multiple-unit stock 1938/9
LMS Royal Saloons Nos 798/9 1941
SR 4-SUB electric multiple-unit 1946
LT 'R' class electric multiple-unit 1952
SR Hastings diesel multiple-unit 1957
Metro-Cammell Pullman car 1960
Prototype Mk II coach No 13252 1963
One of the XP64 train coaches 1964

Also, if still extant in department stock, the original 1954 Derby lightweight diesel multiple-unit cars.

In the descriptions of vehicles which follow these conventions have been used:

(1) All coaches are standard gauge (4ft 8½in) except where shown.

(2) Reading from left to right the description of the coach is given as:

Running number — in the majority of cases — where known: this is the number allocated when built even if the vehicle does not at present display it. In the case of LNER-built stock the pre-1946 number is given first, followed by the later number in brackets.

Type — the class (first/second/third/composite/brake). In some cases a building date is given where coaches of different building dates are included on the same page.

Owner — the present owner of the vehicle, who may not be the operator. The names are given in good faith although some may, in fact, belong to individuals within a company or preservation society.

Location — this is difficult because vehicles may be stored at different locations on the same railway so that in most cases the headquarters of the organisation is given.

Dimensions — where not otherwise shown, the length over the body end panels and width over body panels is given.

The bogie wheelbase is also given in most cases.

Finally, the inclusion of a vehicle in the list is from information used in good faith that it will be preserved and at some date restored.

The details relating to the National Railway Museum/Department of Education & Science exhibits are correct as at June 1975 — noting that some stored vehicles may in time be rotated with those at present on display.

L&MR locomotive Lion with replica coaches.

Just as with the steam locomotive the Liverpool and Manchester Rly — which opened in 1830 — also provided a proving ground for railway coach design. The earliest vehicles were a mixture — some with a horse coach body on a railway underframe. By 1834 at least a reasonable underframe design had been evolved, also a combined spring buffing and drawgear. None of these early vehicles has survived, but the LMS built six replicas — three firsts, three seconds — for the L&MR Centenary Exhibition in 1930.

Of these replicas the firsts have three compartments seating three-a-side with cloth upholstered seats, padded headrests, and a cloth-quilted ceiling. The luggage was carried on the roof — restrained by rails — and box seats were provided on the roof for the guards. The seconds, entirely open, have three compartments seating four-a-side; second-class became closed from the late 1830s. All are four-wheelers and of course without brakes. These coaches were kept at Derby after the 1939-45 war, although one of each type was on display at Clapham.

Type: Firsts (3). **Owner:** Department of Education and Science. **Location:** National Rly Museum, York (1), Liverpool Museum (1 on loan), SGST, Tyseley (1 on loan).

Type: Seconds (3). **Owner:** Department of Education and Science. **Location:** National Rly Museum, York (1), Liverpool Museum (1 on loan), SGST, Tyseley (1 on loan).

Grand Junction Railways
TPO Van (Replica) 1838

The carriage of mails by train goes back to the very earliest days — rail was used for the first time for London to Manchester and Liverpool traffic in 1837, north of Birmingham by rail, southwards by road. By early 1838 a postal sorting van was in use on the Grand Junction railway between Liverpool and Manchester. Soon aferwards John Ramsay of the GPO invented the train and lineside collection/delivery apparatus used on British Railways until 1971. In June 1838 the Grand Junction decided to have built a special mail sorting coach. This was a four-wheeler about 18ft long and weighing some four tons. A pick-up net was provided and there was an iron chute through the bodyside for delivery of the mail pouches. In general outline and mechanical details the sorting coach resembled contemporary passenger stock.

A replica of the GJR sorting coach was built by the LMS in 1938 for an exhibition at Euston Station to commemorate one hundred years' operation of the travelling post office. It was stored at Wolverton until going on display at Clapham.

Type: Travelling Post Office Van. **Owner:** Department of Education and Science. **Location:** National Rly Museum, York.

The replica TPO van.

Bodmin & Wadebridge Railway
Four-wheel Stock

These three vehicles may be the oldest railway passenger coaches in existence in the world. Their date of construction is not certain: both first and second may date from after 1840; the third is probably older. The Bodmin & Wadebridge Railway was an isolated standard gauge line in Cornwall until acquired by the LSWR. When the LSWR main line was extended into Cornwall in 1895, the old B&W coaches were withdrawn but then fortunately saved by Mr W. Panter, the carriage superintendent of the LSWR.

The coaches, all four-wheelers built locally, have a low centre of gravity for the period and are of wooden construction apart from the running gear, which is iron. The third, literally 'open', was probably **no worse than usual for its period: it weighs only 2-3 tons. The** second — which looks like a mobile ship's wheelhouse — sat 16 passengers. The composite, which is very similar to early mainline stock, was originally a first and had cushioned seats. Its body is 13½ft long. The composite coach was last restored in 1960 at Eastleigh and repainted with Prussian blue lower panels, cream uppers and vermilion underframe and window frames. The other coaches are painted blue overall.

The second and third were in the old York museum. The composite was originally on display at Waterloo station and later in store before going to Clapham.

Types: Second (1), Third (1), Composite (1). **Owner:** Department of Education and Science. **Location:** National Rly Museum, York.

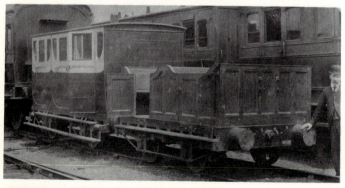

Composite coach (LH) and third (RH) — in LSWR days. LPC.

London & Birmingham Railway
Queen Adelaide's Saloon

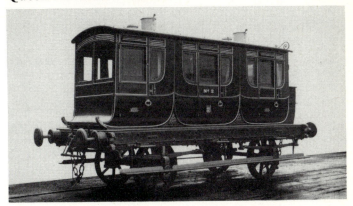

Queen Adelaide's bed carriage.

This bed carriage coach was built for the use of the Dowager Queen Adelaide (wife of William IV) and the body consists of a sleeping compartment, day compartment and coupé. The interior is heavily padded and quilted. A body length of 16ft 6in, body width of 5ft 1½in and interior height of some 5½ft makes for claustrophobia. The body shows some sophistication from the stage-coach outline, as it slopes down at the ends and is also slightly waisted. Large body-end windows are fitted. The exterior is painted in deep claret, which was later the standard livery of the London & Birmingham Railway. There are coats of arms on the bodysides; door and commode handles are gold-plated. Two oil-pots provided illumination. One other feature of interest is that the body was slightly shorter and narrower than the underframe — not unusual at this time. It was originally stored at Wolverton and then on display at Clapham.

Although Queen Adelaide's saloon has happily survived for us to appreciate, two similar four-wheel royal saloons of the 1840s went for scrap within the last 30 years. The oldest, with a centre saloon and two end compartments, was built by the London & Southampton Rly for Queen Victoria in 1844. This later passed to the Shropshire & Montgomeryshire Railway (a light railway) and survived until Army operations in the late 1940s. It was included on

11

the BTC's preserved list in 1956 but subsequently disappeared. The other coach was a coupé built in 1848 by the L & S for Queen Victoria and later passed to the Kent and East Sussex Rly (together with the 1844 coach) via the Plymouth, Devonport & South West Junction Rly. This was bought by the SR from the K&ESR in 1936, but its fate is unknown.

Type: Queen Adelaide's Coach. **Owner:** Department of Education and Science. **Location:** National Rly Museum, York.

William Dargan's Saloon — (5ft 3in gauge)   c1844

This saloon was built for the railway contractor William Dargan by Messrs Dawson of Phibsborough. Dargan presented it to the Midland & Great Western Railway on the completion of his contract for the construction of the Athlone and Galway sections of that railway in 1851. It was then used by the M&GWR as a state carriage until reclassified as a composite in 1904. It was upgraded to first-class only in 1924.

The vehicle is 31ft 3in long. As a first-class coach it seated 17 passengers in two saloons, the smaller one of which led to a lavatory. There is a single door to each bodyside. The interior panelling is in

At Inchicore Works, Dublin, 1964. Herbert Richards.

mahogany. Both underframe and roof were renewed in 1904. Until 1920 the saloon was painted in the M&GWR dark blue livery, being repainted dark lake in 1923. It is now painted dark blue.

During its life as a state carriage it was reputed to be the favourite choice of the Empress Elizabeth of Austria, who patronised the M&GWR.

No: 47. **Type:** William Dargan's Saloon. **Owner:** Belfast Folk & Transport Museum. **Location:** Belfast.

Dublin, Wicklow & Wexford Railway
Third-class Coach — (5ft 3in gauge)   c1844

As received at Belfast Museum. Belfast Transport Museum.

This vehicle was originally built as an open coach for the Dublin & Kingstown Railway. The general, rather spartan appearance is typical of its period for the third-class passenger. The provision of a roof indicates a concession to the Railway Act of 1844, which introduced a series of reforms for third-class passengers. The superstructure is entirely of timber and the interior has plain boarded transverse seats. The coach is 19ft 2in in length. The Dublin & Kingstown — originally 4ft 8½in gauge — later became the Dublin, Wicklow and Wexford, and then the Dublin and South Eastern Railway.

No: 48. **Type:** Third. **Owner:** Belfast Folk & Transport Museum. **Location:** Belfast.

Stockton & Darlington Railway
Composite Coaches
1845?

No. 31 at Stockton.

In the 1835-60 period railway coach design progressed relatively little. Overall length was still around 20ft and the design of the body was a small advance on the 'stage-coach' principles of the earliest days.

Two Stockton & Darlington Railway coaches survive, both first/ second class composites built by Homer & Wilkinson of Darlington. They are very similar, although No 59 has retained more of the original grab rails and guard's roof seat. Handbrakes are fitted: these were worked by the guard from his perch on the roof. The first-class compartment in the centre is well padded and can seat six passengers. The flanking second-class compartments have no upholstery of any kind — and no quarterlight windows. The lighting was by oil pots. As seen in the photograph, rails were fitted on the roof to restrain passengers' luggage. The attractive decorative work on the lower body panels is worth noting.

No 59 was exhibited at the old Queen Street Museum in York. At the time of the S&D centenary celebrations in 1925 No 31 was placed on display on a plinth at Stockton station following a commemorative ceremony. It was removed in 1970.

No: 59. **Type:** Composite (1845). **Owner:** Department of Education and Science. **Location:** National Rly Museum, York.
No: 31. **Type:** Composite (1846). **Owner:** Department of Education and Science. **Location:** Darlington North Road Museum (on loan).

Eastern Counties Railway
Four-wheel Coach   1851

The idea of railway coaches having an economic life of 30-40 years obviously dated back to the earlier days. This Eastern Counties railway coach was withdrawn by the GER in 1901 when running as an inspection saloon. It was originally a three-compartment first. The body is 17ft 6in long and again, as with the earlier coaches, the interior height is just under six feet. Tare weight is eight tons.

This vehicle was fitted with end windows on conversion to an inspection saloon and the interior changed. After withdrawal it was stored in the GER's Stratford Carriage Works paint-shop and survived wartime bombing. Not on display before going to Bressingham. However, restoration has been shelved as the museum has no details of the original livery.

Type: First. **Owner:** Department of Education and Science. **Location:** Bressingham Museum (on loan).

Eastern Counties three-compartment first, believed to be of the same type as the preserved vehicle.

Festiniog Railway
Four-wheel Stock — (1ft 11⅝in gauge)   1864-7

These little coaches (Nos 3-7) with their knifeboard (back-to-back) wooden seats are an intriguing survival from the pioneering Festiniog narrow-gauge railway. They were in service more or less regularly until the cessation of passenger services (under the original company's operation) in 1939. Since the reopening under the new régime in 1955 (although not used at first) they have continued to see service, usually formed in a relief set. They were thoroughly overhauled from the early 1970s.

Fourteen coaches were built from 1864-7 by Brown Marshalls and Ashbury Rly Carr & Iron Co Ltd to four different types. Six remained in 1939; one was subsequently scrapped. Nos 3, 4, 5 (B. Marshalls) are closed and glazed coaches, 10ft 3in long and 6ft 8in in width — probably originally first, second and third class respectively. No 6 (B. Marshalls), although to the same dimensions, is open above the waist with wire screens to the waist. No 7 (B. Marshalls) was, after 1955, thought to have been the same as No 6 but more recently found to have been completely open and has been restored in this condition. All five coaches are vacuum-piped only, ie without through brakes.

The quarrymen's coaches originally totalled 44 (in 1875), for use on the services for slate quarry and other workers. The last of these services also ran in 1939. Two, probably dating from the 1870s, have survived in recognisable condition. No 8, originally matchboarded, has been restored as a quarrymen's coach and works as a vacuum-

Four-wheel stock (left to right): Van No 2, coach No 5 and two of the 3-5 series. Taken 1959.

braked guard's van with the other four-wheelers. No 2 van (originally a quarrymen's coach) is normally used for engineer's works trains.

Nos: 3, 4, 5. **Type:** Third class 'knifeboard' coach. **Owner:** Festiniog Rly. **Location:** Porthmadog.
Nos: 6, 7. **Type:** Third class 'knifeboard' open coach. **Owner:** Festiniog Rly **Location:** Porthmadog.
No: 8. **Type:** Quarrymen's coach (187?). **Owner:** Festiniog Rly. **Location:** Porthmadog.
No: 2. **Type:** Van (rebuilt). **Owner:** Festiniog Rly. **Location:** Porthmadog.

Talyllyn Railway
Four-wheel Stock (2ft 3in gauge)   1866

The Talyllyn was opened for freight in late 1865 and for passengers in December 1866. The original coaches were Nos 1-3 and the brake van No 5. These and No 4, built 1870, formed the only passenger stock on the railway until 1952 — although not originally numbered. Nos 1-3 are very simple vehicles with heavy oak frames, iron wheel centres and running gear, and wooden bodies. They have three compartments with no interior partitions, except for No 3, which has one partition providing a first or second-class compartment. The doors on the non-platform side have always been sealed due to limited clearances. All were built by Brown, Marshalls. Nos 1/2 are 14ft 6in long, No 3 one foot shorter and 5ft 3½in wide. The brake van, although on the same running gear as the coaches, has sliding doors on each side and, from 1900, has been used to issue tickets to

Left to right: No 4, either Nos 1/2 or 3 and brake van. B. A. Butt.

17

passengers joining at intermediate stations. A ticket window was cut in the platform side lookout. The body has outside framing over boarded panelling. No 4, built by Lancaster Wagon Co, is of similar dimensions to the other vehicles but different in appearance. It originally had outside body framing with half-height doors, but in the 1920s was rebuilt with full-height doors and additional body planking. The running gear was originally fixed to sub-frames but deteriorated to the extent that a new steel underframe was fitted in 1958. All are painted a crimson red with brown ends and body mouldings, lined out in green on the side panels.

Nos: 1, 2. **Type:** Third (1867). **Owner:** Talyllyn Rly. **Location:** Towyn.
No: 3. **Type:** Third (1866). **Owner:** Talyllyn Rly. **Location:** Towyn.
No: 4. **Type:** Third (c1870). **Owner:** Talyllyn Rly. **Location:** Towyn.
No: 5. **Type:** Brake van (1866). **Owner:** Talyllyn Rly. **Location:** Towyn.

Stockton & Darlington Railway
Four-wheel Coach   1867

This coach, a four-wheeled, four-compartment vehicle, was built at the S&D carriage works at Darlington. As will be noted from the photograph, the body is of horizontal planking with outside framing — a type of construction often used at the time. The running gear is simple, with large spoked wheels. By the time the coach appeared,

Being restored at Shildon. 1975.

the S&D had already been formally amalgamated with the North Eastern Railway — as from July 1863 — but continued as a separate concern until 1873. The NER sold the coach to the Forcett Limestone Company in 1884 (the Forcett Railway was a branch from the Darlington to Barnard Castle line west of Darlington). The coach was found many years later in Forcett station and was then exhibited for some time at York Railway Museum. In the 1960s it was removed for storage and more recently was transferred to the care of the North of England Open Air Museum. By this time it was in poor condition. Fortunately the works manager at the Shildon works of British Rail Engineering Ltd was prepared to take on the job of restoration and some of the works staff volunteered to restore the coach in their leisure time. After four months' work it was outshopped in March 1975. Quite a lot of replacement parts were manufactured at Shildon and lacquers and paints were donated by a local firm. The interior has been finished in grained oak and the exterior in teak.

No: 179. **Type:** Third. **Owner:** North of England Open Air Museum. **Location:** Beamish (in store).

London & North Western Railway
Queen Victoria's Saloon   1869

Queen Victoria was probably one of the first long-distance railway travellers and on the long journeys to Deeside (Balmoral) required extensive facilities. A four-wheel coach was built for Queen Victoria in 1861 by the LNWR and this was replaced in 1869 by two Wolverton-built six-wheelers. These inherited some of the furnishings of the 1861 car. Each of the later cars was 30ft long and they were close-coupled, connected by a closed flexible gangway. One car was for day use with a sofa, chairs, lavatories and an attendant's compartment. The other car was the night saloon with two beds in the main saloon. There was also a dresser's compartment. The floors were double and filled with cork and the cars had double roofs. The interior wall and ceiling covering was ornate, high-Victorian quilted or watered silk. Lighting was by conventional oil pots and candles (later gas). Footwarmers provided the only heating in the early days. The coaches were first used on 14 May 1869 by Queen Victoria between Windsor and Ballater (later Balmoral).

   Except for the replacement of the outer rigid axle on each coach

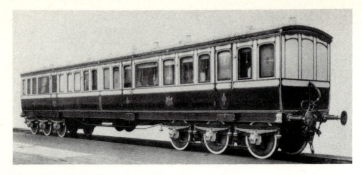

Queen Victoria's saloon after 1895.

by a radial axle in 1890, the coaches remained much the same until 1895. Then the two bodies were combined on a new 60ft underframe but in deference to the Queen's wishes few other changes were made. Electric lighting was installed, but both oil lamps and candles were retained! Six-wheel, 11ft 6in wheelbase bogies were fitted. From the start the coach(es) were painted with claret lower panels and lined-out white above. The Queen used the saloon for the last time on 6/7 November 1900 between Ballater and Windsor. The saloon was stored for many years at Wolverton before going to Clapham in 1960/1. It was exhibited at the Stockton & Darlington Railway Centenary celebrations in 1925.

Type: Queen Victoria's Saloon. **Owner:** Department of Education and Science. **Location:** National Rly Museum, York.

North Eastern Railway
Saloons 1870/1

Saloon No 1173 has the distinction of serving the North Eastern Railway and its successors for 97 years, being withdrawn at York in 1967. It is of conventional outline for its time, measures 29ft 10in over the buffers (which are of the old wood-padded type), weighs 11½ tons and was originally built as a first-class passenger saloon. It was renovated in 1892 and later allocated as an inspection saloon for the use of the superintendent of the line, later the general traffic manager. In this condtion, and at present, it has a saloon for ten persons at one end and at the other end a kitchen and an attendant's compartment. As LNER 21173 it was allocated to the district engineer Hull until moving to York in 1963 (by then, E 900270E) for use in the chief civil engineer's gauging train. In short, a remarkable survivor, particularly as it has been altered so little.

Saloon No 1661 is an unusual slab-sided clerestory roof coach built by the Stockton & Darlington Railway as a six-wheel third-class saloon. It came to the NER in 1876 and was converted to an inspection saloon in 1884 for use of the locomotive superintendent at Gateshead Works, later at Darlington. It assumed its present form in 1904 when rebuilt as a 40ft bogie vehicle and continued to be used by the locomotive superintendent, working with the 2-2-4T locomotive *Aerolite* — now preserved at York. The main saloon, which has four end windows, seats 15 (originally 19). There are a kitchen, lavatory and attendant's compartment at the other end. Over the years few changes were made to the saloon, apart from the fitting of new windows. Later LNER 21661, it then became BR E 902179E and was based at York until withdrawn in 1969. More unusually, it became a film star on the KWVR when used as 'The Old Gentleman's Carriage' in the production of *The Railway Children.* It is now finished in varnished teak as LNE 21661.

No: 1173. **Type:** Inspection Saloon (1870). **Owner:** Keighley & Worth Valley Soc. **Location:** Haworth.
No: 1661. **Type:** Inspection Saloon (1871). **Owner:** Private. **Location:** Haworth.

No 1173 as purchased for preservation. A. Cox.

Coach No. 17 (built 1876) as restored.

The Festiniog Railway deserves a special place in this record of historic coaches, in particular for its bogie stock built from 1871 onwards, as these were probably the first bogie coaches in service in the British Isles. Until the 1880s the FR was a pace-setter in railway practice. The earliest coaches, Nos 15/6, were built in 1871 — bodies by Brown Marshalls — and were designed by the FR's engineer, Charles Spooner. They have wrought-iron underframes and body framing. At each end they have platforms to enable passengers to cross between the coaches. Also at each end is a small compartment originally providing additional seating. Each coach was originally a tri-composite, later with one first-class compartment only, with padded upholstery. The other compartment partitions are half-height only, originally with oil lighting. Both coaches were left to the elements after 1939 but have now been fully rebuilt — No 15 in 1960, No 16 in 1969. Both are now painted brick red. Their length is 35ft 9in.

Nos 17-20, also designed by Spooner, have iron underframes but with timber framing and panels. The bodies of Nos 17/18 were built by Brown Marshalls, those of Nos 19/20 by Gloucester Wagon. All are lighter and shorter (Nos 17/8, 32ft, Nos 19/20 34ft) than Nos 15/6. The bodies have tumblehome sides. All were originally tri-composites — later 1st/3rd only — but with differing compartment layouts. All were in service by 1963.

Nos 10-12 were originally bogie luggage vans Nos 2/4/5 built by the FR Boston Lodge Works. All were rebuilt as brake coaches 1921-30. Nos 11/12 were rebuilt after 1955; the first as an observation saloon and No 12 as a third brake buffet. Both are 24ft 6in long. No 10 existed as a third brake before 1955. More recently it was used for service purposes only and now awaits rebuilding.

No: 10. **Type:** Third Brake, 2 compartments (1872). **Owner:** Festiniog Rly. **Location:** Porthmadog.
Nos: 15, 16. **Type:** 1st/3rd Composite (1871). **Owner:** Festiniog Rly. **Location:** Porthmadog.
Nos: 17, 18. **Type:** 1st/3rd Composite (1876). **Owner:** Festiniog Rly. **Location:** Porthmadog.
Nos: 19, 20. **Type:** 1st/3rd Composite (1879). **Owner:** Festiniog Rly. **Location:** Porthmadog.
No: 11. **Type:** 1st class observation saloon (1880). **Owner:** Festiniog Rly. **Location:** Porthmadog.
No: 12. **Type:** 3rd Saloon buffet (1880). **Owner:** Festiniog Rly. **Location:** Porthmadog.

North London Railway
Directors' Saloon   1872

This modest little directorial saloon, some 27ft in length, is interesting not only as being representative of a small railway but of the North London's standard stock. The standard North London Railway coach was introduced in the 1860s: a teak-bodied, four-wheeler of which eight to twelve would be close-coupled in sets. These coaches were the first — and last — to have coal gas lighting. From 1916 they were displaced by LNWR electric multiple-units, but survived

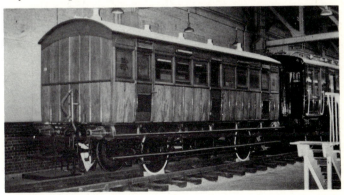

No 1032 on display at Clapham.

into the 1930s on through services from Broad Street to the North London suburbs. Many were sold off to small railways round the country. The body of one has been preserved by the Lakeside Railway Society at Haverthwaite. The director's saloon is in excellent condition with a varnished teak body, typical NLR gas lamp fittings and distinguished by etched and ground glass windows. It was stored at Lostock Hall, Preston, before display at Clapham.

No: 1032. **Type:** Director's Saloon. **Owner:** Department of Education & Science. **Location:** National Rly Museum, York.

Midland Railway
'Midland' Pullman Car   1874

A Pullman drawing-room car, as built.

One of the biggest advances in rolling stock design resulted from the import of American Pullman sleeping and day cars to Britain from 1874. This was the result of the Midland Railway's general manager, James Allport, visiting the United States in 1872. As a consequence, at their own expense, the Pullman Palace Car Company brought over the crated parts of a sleeping car, to be named *Midland*, for assembly at the Midland Railway's Derby works. This car, 58ft 5in over the end platforms, 9ft wide and mounted on American-type wooden bogies, was completed in January 1874. The interior had 24 upper and lower sleeping berths, with attendant seats.

In accordance with American Pullman practice there was a central gangway. The body and underframe were a single structure, all timber but reinforced by iron pillars, carlines and truss-rods. The open end platforms had wrought iron gates; above was the hooded roof end and, overall, the coach had a typical American clerestory. Externally, the coaches were painted a rich chocolate mahogany

24

brown with gold leaf decoration. The interior woodwork was American walnut offset by quilted oil cloth for the ceilings. Apart from paraffin oil lamps, passenger comfort was provided by an oil-fired heating system individual to each car. The Midland Pullmans first went into service — as a complete train — in June 1874 between St Pancras and Bradford.

These original Pullmans were taken into MR stock in 1888 and continued as sleeping cars. Even then, they were not finished and were used as saloons or on push-and-pull trains into the 1900s. Finally they were grounded to serve as sheds and mess rooms — a tribute to their very solid construction. Two survived in the 1960s/1970s — one at Skipton ('Midland') and another at Hellifield. So, although the Midland Railway project has saved what is virtually a grounded hulk, we now have at least a tangible reminder of an American Pullman.

Type: 'Midland' sleeping car (body only). **Owner:** Derby Corporation Museum (Midland Railway Project). **Location:** Butterley.

Maryport & Carlisle Railway
Six-wheel Third   c1875

This coach, a six-wheel third built by Birmingham RCW, is again typical of its period: simple in construction and with plain bodyside mouldings. As purchased by the RPS it was not fitted with continuous brakes. What is interesting is that it is one of the very few survivors from one of the smaller pre-Grouping railways. The Maryport & Carlisle had a route mileage of some 42 miles only and

As restored 1975. Iain Smith.

relied on iron ore and coal traffic for its comparatively flourishing prosperity. In 1922, before Grouping, it had only 71 passenger vehicles, most of which were four-wheeled. The coach preserved at Chasewater has been restored to the M&C post-1905 livery of dark green lower and greenish-white upper panels with gold lining-out.

The coach has had an interesting history, being sold, possibly in the 1930s, to the Cannock and Rugeley Colliery Co at Rawnsley for the miners' service to Hednesford (Staffs). By 1955 it was out of use so that its subsequent restoration has meant a lot of hard work.

No: 11. **Type:** Third. **Owner:** RPS, Chasewater Lt Rly. **Location:** Brownhills.

MSLR
Four Six-wheel Stock   1876

Up until the late 1870s the MSLR followed a very conservative line in coach design until it was spurred on by the Midland to do better. The MSLR also took its time in producing bogie stock which it initially considered as unlikely to make any headway. The five coaches which have been preserved are of largely the same design: arc-roofed bodies with characteristic half-round tops to the quarterlights (windows) and sharply radiused corners to the body mouldings. Many of these coaches, some with flat flame gas lights, were in service on the LNER in the mid-1930s.

No 103 (later LNER 5176) has two third-class compartments, one second and one first, with a central luggage compartment. Latterly a stores van at Retford, it came to Haworth in 1965 and is being

No 103 as preserved. R. Higgins.

26

thoroughly restored. No 373, a five-compartment third, is of similar design. No 1076, a five-compartment third, was built for the Manchester South Junction & Altrincham Joint services and is 34ft 8½in in length. It was converted as a camping coach in the 1930s. The original identity of the other coach at Quainton Road is still in doubt: it was taken into departmental stock at Hull in 1935. The six-wheeler at Chasewater is being restored extensively. It has one first-and two second-class compartments. It was sold to the Easingwold Light Railway in 1946 and used until 1956.

No: 103. **Type:** Four-wheel Tri Compo Bk. (1876). **Owner:** Vintage Carriages Trust. **Location:** Haworth.
No: 1076. **Type:** Six-wheel Third (1890). **Owner:** GCR Coach Group. **Location:** Quainton Road.
No: DE 320256. **Type:** Six-wheel Third (1890). **Owner:** GCR Coach Group. **Location:** Quainton Road.
No: 1470? **Type:** Six-wheel Brake Composite (c1890). **Owner:** RPS, Chasewater Lt Rly. **Location:** Brownhills.

Isle of Man Railway
Passenger Coaches (3ft gauge)   1876-1926

It would be impossible to describe adequately the remaining passenger coaches of the IOMR in such a short space. Instead the reader is referred to the excellent *The Isle of Man Railway,* by J. I. C. Boyd (published by Oakwood Press 1973). At present the future

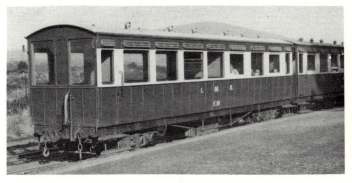

F30. T.K. Widd.

F18 and F34. R.A. Panting.

of the island's railway is problematic, so that what will eventually happen to the coaches is anyone's guess. The present list is based on those vehicles extant in 1973.

No	Builder	Date	Type	Body length x width
F1-F4	Brown Marshall	1876	Third Brake	35ft x 7ft
F5/6	Brown Marshall	1876	Compo Brake	35ft x 7ft
F7	Ashbury	1881	Compo Brake	35ft x 7ft
F9-12	Brown Marshall	1881	Third	35ft x 7ft
F13-5	Brown Marshall	1894	Compo Brake	35ft x 7ft
F16-8	Brown Marshall	1894	Third Brake	35ft x 7ft
F19	Brown Marshall	1894	Third Brake	35ft x 7ft
F20	Metropolitan RCW	1896	Third Brake	35ft x 7ft
F21-4	Metropolitan RCW	1896	Compo Brake	35ft 11in x 7ft
F25/6	Metropolitan RCW	1896	Third Brake	35ft 11in x 7ft
F27/8	Metropolitan RCW	1897	Passenger Bk Van	35ft x 7ft
F29-32	Metropolitan RCW	1905	Saloon Third	36ft 11in x 7ft
F33/4	Metropolitan RCW	1905	Third Brake	37ft x 7ft
F35/6	Metropolitan RCW	1905	Saloon Compo	36ft 11in x 7ft
F37/8 (ex MNR)	Hurst Nelson	1899	Compo Brake	35ft 6in x 7ft 0¾in
F39 (ex MNR)	Bristol & S Wales Carr	1887	Compo Brake	30ft x 7ft 2in
F40-4	Metropolitan RCW	1907/8	Compo Brake	37ft x 7ft
F45/6	Metropolitan RCW	1913	Compo Brake	36ft 11in x 7ft
F47/8	Metropolitan RCW	1923	Third	36ft 11in x 7ft
F49	Metropolitan RCW	1926	Third Brake	37ft x 7ft
F50/7/63-7/70/1	rebuilt 4-whl stock		Third	33ft 2in x 7ft
F54	rebuilt 4-whl stock		Third Brake	34ft 2in x 7ft
F62/8/73/4	rebuilt 4-whl stock		Third	34ft 2in x 7ft
F75	rebuilt 4-whl stock		Saloon Third	34ft 2in x 7ft

F1-6 are the original compartment stock, F7-12 being similar. F13-9 are a later series with 'tumblehome' bodysides as are Nos F20-6. F27/8 are full-length brakes known as the 'Empress Vans'. F29-32/5/6 are saloon coaches, the first Isle of Man Railway stock with steel underframes. Entrance is by doors at each end. F33/4, 40-9 are compartment coaches. F37-9 are ex-Manx Northern Railway coaches — all compartment stock. F50-75 were constructed by mounting pairs of four-wheel coach bodies built in 1873/4 on new bogie underframes during 1909-26. F75 has the bodies from two special saloons. Nearly all are painted in the standard cream-and-red livery. Some are part derelict. There are a number of very disused Manx Northern Railway six-wheelers still in existence.

Great Eastern Railway
Four/Six-wheel Stock   1878

The GER took some time to live down its early poor reputation for passenger comfort. James Holden went to the GER as Locomotive and Carriage Superintendent in 1885 from Swindon, where the standard of rolling stock design was of a much higher quality. From the 1890s GER passenger coaches improved steadily in specification. The first No 19, at Chappel, was brought from Felixstowe where it had been used as a holiday home: only the body with solebars remains. No 19's body is 27ft long with four compartments. It was built by Birmingham RCW for working in close-coupled London suburban sets. In 1903 the body was widened by 1ft to provide five-a-side seating and the coach was withdrawn in 1913. The body is in surprisingly good condition and appropriate wheelsets and under-frame equipment are being obtained so that it may run again. No 553, a passenger brake van, is virtually complete. It was built for main line use at Stratford Works. The brake end, with prominent guard's lookouts, has the characteristic end windows. The under-frame is of wood and iron and the Westinghouse brake gear still survives. The bodyside panelling is largely intact. It was withdrawn

GER brake No 553 with MS&L 373. G.D. King.

in 1934 as LNER 6376 and became a tool van. The body of a third-class family saloon is preserved at Audley End — used as club headquarters by the Saffron Walden Model Railway Club.

No: 19. **Type:** Four-wheel First (body only). (1878). **Owner:** Stour V RPS. **Location:** Chappel (Essex).

No: 553. **Type:** Six-wheel Pass Bk Van (1890). **Owner:** GER Group. **Location:** Chappel (Essex).

No: 44. **Type:** Six-wheel Pass Bk Van (1885). **Owner:** RPS, Chasewater Lt Rly. **Location:** Brownhills.

Lancashire and Yorkshire Railways
Saloon   1878

This unusual vehicle with balconies at each end was built as a six-wheeler at the L&YR's Miles Platting works, as Director's saloon No 1. In 1908 it was mounted on short wheelbase bogies and rebuilt at Newton Heath carriage works, becoming inspection saloon No 2 after replacement by a new director's saloon, now preserved at Sheringham (page 88). Subsequently it was used as an engineer's saloon in Lancashire until 1948, when it was transferred to Inverness. It had previously been LMS No 10772 and latterly BR No SC 45038. There is a saloon at each end and a kitchen and lavatory in the centre. The KWVRPS intend to restore the coach to L&YR livery.

No: 2. **Type:** Inspection Saloon. **Owner:** Keighley & Worth Valley RPS. **Location:** Haworth.

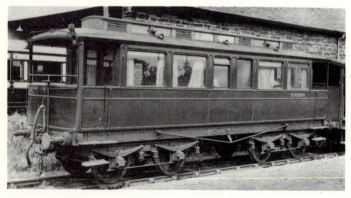

The saloon as SC 45038M, 1962. T.J. Edgington.

LCDR
Four/Six-wheel Stock

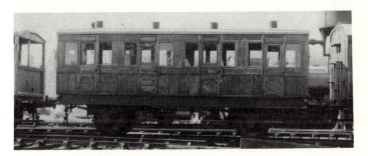

First-class LCDR saloon as restored, July 1975. J.D. Parsons.

These two coaches exemplify the backwardness of the railways in South-East England in the 1880s, with but little advance over practice in the 1870s. The first, built in 1880 at the LCDR Longhedge works, has four compartments and is distinguished — if that is the word — by austere square-cornered windows and vertical lower-body panels. The recessed oval panels in the door are an unusual feature. The coach is remarkably complete with body-end steps for access to the roof and grab rail still intact. It is being restored to varnished teak finish. Purchased in 1962 by the former London Railway Preservation Society, the coach was used in restored state for the first time in August 1973. The other coach, a brake third, is similarly frugal in specification with square-cornered windows and body mouldings. Its original LCDR running number is not known, but it became SECR No 2781 and later SR No 3630, passing to service stock in 1935. It was purchased from Redbridge (Southampton) early in 1962 and is as yet unrestored.

Type: Four-wheel First.**Owner:** Quainton Rly Soc. **Location:** Quainton Road. **No:** SECR 2781. **Type:** Six-wheel Third Brake (1894). **Owner:** Bluebell RPS. **Location:** Sheffield Park.

Midland Railway
Four/Six-wheel Stock

The background to Midland Railway coaching stock practice is of relevance. From 1875 the company abolished second class, at the

The six-wheeled composite at Haworth. R. Higgins.

same time upholstering third class accommodation and generally improving it. After 1874 no more four-wheel coaching stock was built except for vans, such as that at Brownhills. Between 1877 and 1892 the clerestory roof was discarded in favour of a plain arc roof.

The three six-wheelers are typical of Midland practice from the late 1870s until the 1890s. They have the plain arc roof with full bodyside panelling. The composite at Haworth, 34ft long and 8ft 6in wide, is of a numerous type and has two first- and two third-class compartments with a central luggage compartment. It still has body-end steps and handrail. This coach was used in departmental stock until 1968 and awaits restoration. No 7243 is a third brake acquired from the British Steel Corporation's Shotton works in 1972 and has not yet been restored. No 901 is in fully restored condition.

Of the vans, that at Brownhills is in generally good condition — an arc-roofed four-wheeler. Much of the exterior panelling is still intact as are the various external fittings. However, on one side there is a large window put in by the Manchester Ship Canal Co, who used it as a mobile pay office. The two six-wheeled vans preserved for the Midland Railway project at Butterley are of one type. They have clerestory roofs and follow Thomas Clayton's design introduced in 1897 with square-cornered panelling and the clerestory end-boarding integral with the ends. Both are largely complete but await extensive restoration, having been latterly engineers' vehicles.

Type: Four-wheel Passenger Brake Van (c1880). **Owner:** RP Soc. **Location:** Brownhills.

No: 7243. **Type:** Six-wheel Third Brake (1884). **Owner:** Derby Corporation. **Location:** Butterley.

No: 901. **Type:** Six-wheel Luggage Composite (1885). **Owner:** Department of Education & Science. **Location:** National Rly Museum.

No: (DM 284677). **Type:** Six-wheel Luggage Composite (1886). **Owner:** Vintage Carriages Trust. **Location:** Haworth.

No: 485. **Type:** Six-wheel Passenger Brake Van (1898). **Owner:** Derby Corporation. **Location:** Butterley.

No: (DM 198587). **Type:** Six-wheel Passenger Brake Vans (1902). **Owner:** Derby Corporation. **Location:** Butterley.

Great Western Railway
Saloon   1881

This saloon is the earliest GWR coach in existence and externally follows the bodyside design common up to the mid-1880s, with deep cantrail panels (above the windows) and heavy beading around the large saloon windows. The coach is 46ft 6¾in long and 9ft 0¾in wide, running on 8ft 6in wheelbase Dean suspension bogies. The clerestory is also worth study as of the earliest GWR type with a single arc to each deck. No 241 was built as a first-class family saloon — for hire to any family for their exlusive use for journeys to destinations on or off the GWR system. The original accommodation

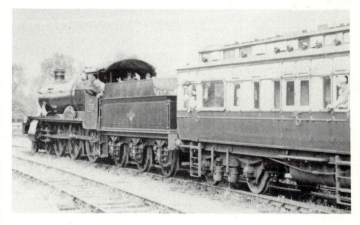

The coach at Ashchurch. D.A. Anderson.

consisted of a saloon at each end with facilities for servants and the lavatories in the centre of the body. After the 1907 renumbering the coach became No 9044 (in the saloon series) and was the 'special saloon — for hire to any family for their exclusive use for journeys to District Engineer's saloon, used first at Wolverhampton and later at Oswestry and Shrewsbury. The saloon was numbered 80973 in departmental stock and by late 1962 was lying at Swindon works awaiting scrapping. Fortunately David Rouse — who deserves special praise for his hard work in 'saving' historic coaches — stepped in to raise £225 to purchase No 241, which then entered honourable retirement at Ashchurch.

No: 241. **Type:** First Saloon. **Owner:** Dowty RPS. **Location:** Ashchurch.

Penrhyn Railway
1ft 10¾in gauge Stock   1882

The Penrhyn Railway was the earliest narrow-gauge railway in North Wales, being opened in 1801 from Penrhyn Slate Quarry to Port Penrhyn. It was relaid on a new route in 1876-7 when steam traction

Lord Penrhyns Saloon. National Trust.

was introduced. All rail facilities were withdrawn in 1962. Various items of rolling stock were placed on permanent loan with the National Trust by Penrhyn Quarries Ltd in 1963.

The saloon was probably built in 1882 for the use of the then quarries' owner, Lord Penrhyn, and his agent on trips between the port and the quarries. The builder is unknown. It is a small four-wheeler, 11ft 10in in length and 5ft 6in wide, which can seat 15 persons and is upholstered in blue cloth. It is painted in cream and black. The saloon was stored out of use for some time before the closure of the railways.

The Quarrymen's coach is one of many simple four-wheeled vehicles used to take workers to and from the quarries, in trains arranged and paid for by the men. The last quarrymen's train ran in 1951. The coach can seat 24 on simple wooden benches without doors. No brakes are fitted.

The Official's car, seating six, was used within the quarries. It was probably built in the late nineteenth century. Of simple wooden construction, the seats resemble a set of three church pews.

Type: Lord Penrhyn's Saloon. **Owner:** National Trust. **Location:** Penrhyn Castle.
Type: Quarrymen's Open Coach. **Owner:** National Trust. **Location:** Penrhyn Castle.
Type: Quarries' Officials' Car. **Owner:** National Trust. **Location:** Penrhyn Castle.

West Donegal Railway
Saloon (3ft Gauge) 1882

Saloon No 1 is the only survivor of the original rolling stock of the West Donegal Railway, later the County Donegal Railways Joint Committee, which closed completely in 1960. It is a six-wheeler with a panelled wooden body 31ft in length and was built by the Railway Carriage & Wagon Co. Originally a tri-compo, it was later converted to a first-class saloon with end windows. The interior consists of three saloons, the centre one having longitudinal bench seats and a lavatory cubicle. In all there is seating for 32 and the lighting is by acetylene. No 1 survived and was available for private hire traffic until the end of the County Donegal in 1960. Other passenger stock was purchased for preservation and stored on the railway, but most has now been vandalised beyond redemption.

No: 1. **Type:** Saloon. **Owner:** Belfast Transport Museum. **Location:** Belfast.

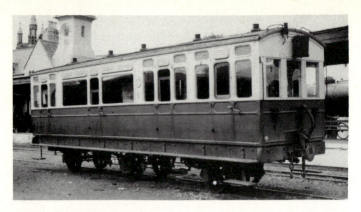

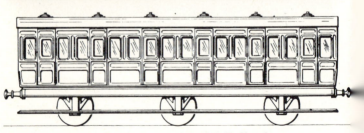

Top: *West Donegal Saloon No 1 Stranolar, 1959*. F. Church.
Centre: *Wisbech & Upwell Tramway No 8 in original condition*. LPC.
Bottom: *Outline diagram of NBR six-wheel coach*.

Wisbech & Upwell Tramway
Saloon Coach   1884

Roadside tramways were rare in Britain and now they have all gone:
the Wisbech & Upwell ran for the last time in 1966. This line ran
through the Fens, and a passenger service was operated until 1927.
There were six four-wheel coaches with balconies and two bogie
vehicles. The latter, Nos 7 (composite) and 8 (third), were 37ft 4½in
over the buffers, 8ft wide and ran on 4ft 6in wheelbase bogies. The
body with an arc roof and lattice balconies was low-slung — only 3ft
above rail level. The saloon interior was similar to a tram. After
withdrawal of the passenger services the bogie coaches were transferred to the Kelvedon & Tollesbury Light Railway, where they
worked until 1950, when that line closed to passengers. No 8, by
then numbered E 60462E, was kept for preservation and used in the
making of the Ealing Studios' comedy, *The Titfield Thunderbolt*.
This reprieve was short-lived as Stratford Works scrapped the coach
in the late 1950s — together with a carefully restored LT & SR third
of 1910. What vandalism! Early in 1974 the body of Wisbech saloon
No 8 was discovered in use as a store in the Fens and is to be
restored at Cambridge.

No: 8. **Type:** Third Saloon. **Owner:** Cambridge Museum of Technology
Trust. **Location:** Cambridge (not on show).

North British Railway
Six-wheel Coach   c1885

This coach has recently been acquired by the SRPS from departmental use at Edinburgh Waverley. It is 35ft 6½in long and 7ft 6in
wide. The coach has six compartments, each 5ft 9 in long. As built
the tare weight was just under 13½ tons.

No: not known (DE 773090). **Type:** Third. **Owner:** Scottish RPS. **Location:**
Falkirk.

WCJS
TPO Coach   1885

It took some time for bogie coaches to be generally introduced in
Britain. Some rigid eight-wheelers were produced by railways such

No 186 as restored.

as the Metropolitan. One other solution lay in the adoption of radial trucks, where the inner of the four axles were rigid and the outer ones mounted in a frame with movement controlled by curved guide-blocks and lateral play restrained by springs. This arrangement was used by the LNWR to the design evolved by the locomotive and carriage superintendents, Francis Webb and Richard Bore, respectively. The overall length of coaches built with radial trucks was 42ft. Ordinary coaches, sleeping cars and TPOs were constructed with this feature. No 186, a postal sorting van, is the only Webb radial truck vehicle preserved. It has fairly flat bodysides with a three-centre roof and offset passenger gangways. Mail-bag pick-up and delivery apparatus is preserved intact. It is finished in the standard West Coast Joint Stock/LNWR livery of off-white upper and deep lake lower panels, fully lined out. No 186 was built for the predecessor of the 'West Coast Postal' on the London-Aberdeen service which started in July 1885; the first time connecting gangways were used. It was exhibited at the TPO centenary celebrations in 1938, later stored at Lostock Hall and more recently on display at Clapham.

No: 186. **Type:** Postal Sorting Van. **Owner:** Department of Education & Science. **Location:** National Rly Museum, York.

Cavan & Leitrim Railway
Clerestory Saloon (3ft gauge) 1887

The Cavan & Leitrim Railway was incorporated under the Tramways (Ireland) Act of 1883 and the main line from Dromod to Belturbet was opened to passenger traffic in October 1887. Twenty-four clerestory-roofed bogie coaches were supplied to the C&L by the Metropolitan Carriage & Wagon Co. These all had open-end platforms, longitudinal seating, vacuum brakes and centre buffers.

No 6, 40ft in length, was originally a composite with individual armchairs in the first-class compartment, later replaced by bench seats. Still later the coach was modified as a third brake and had double luggage doors fitted to the first-class saloon, which became the guard's compartment; a stove was then provided for the guard's use.

In 1925 the C&L became part of the Great Southern Railway, later of Coras Iompair Eireann. The railway survived to be Ireland's last exclusively steam-operated narrow-gauge line and closed down completely on 1 April 1959.

No 6 is now undergoing extensive restoration, being returned to its composite form with bench seats. It will be repainted dark maroon.

No: 6. **Type:** Composite. **Owner:** Belfast Folk & Transport Museum. **Location:** Belfast.

C&L composite brake. T.K. Widd.

Great Western Railway
Coach No 820

1887

GWR composite No 6244 as converted to a camping coach; it was similar to No 820.

This six-wheeler clerestory tri-composite was built in 1887 to diagram U29 and has the distinctive feature of deep panels above the windows. It was one of the many built as 'convertibles', with standard gauge width body mounted on a broad-gauge underframe. It was in traffic until about 1934 (by then numbered 6820), emerging early the next year as a camping coach with the gaslight fittings removed. It was numbered 9962 and was one of a total of 65 four and six-wheel coaches converted between 1931 and 1935 for camping holidays. During the 1950s many of these conversions enjoyed a further extension of life as service vehicles, which ensured the survival of No 820. The coach is as yet unrestored and there are no plans at present for its future display.

No: 820. **Type:** Tri-composite (six-wheel). **Owner:** Department of Education & Science. **Location:** On loan to Bristol City Museum (in store).

Great Northern Railway
Six-wheel Stock

1887

The Great Northern passenger stock of the 1880s and 1890s had a distinctive outline and finish. The roof was a flattened semi-ellipse and the body panelled in varnished teak with rectangular mouldings and square-cornered windows. The running gear was of no particular interest, although the centre axle was given free play. Despite building a bogie coach in 1874, the GNR stuck to six-wheelers into the 1890s both for its own services and for the East Coast Joint Stock used on Anglo-Scottish services. Among the many coaches built to the standard GNR six-wheeler outline was the first side-corridor ordinary coach for a British railway, completed in 1882. Such vehicles were displaced by modern bogie stock in main line service from the 1900s, but being remarkably tough continued to give reliable if uncomfortable service into the 1940s. When Stirling 4-2-2 No 1 was brought out of York Museum in 1938 there was no difficulty in gathering a representative train of old GN/ECJS six-wheelers to accompany it. Others had been combined on new frames as articulated twins from 1907 onwards. Many survived in service stock until the 1960s — not bad going!

Of the survivors, that at Haworth (original number unknown) is a four-compartment brake third with characteristic GNR guard's lookouts. It was taken out of service stock c 1964/5. No 1470, also a four-compartment brake third, is 34ft 10½in long and 8ft 10¾in wide. It survived in service stock at Hitchin until 1963 and was purchased in 1969. Now nicely restored to varnished teal finish, it is usually paired with the LCDR coach (page 31) on open-day specials. No 948 has been fully restored.

No: 948. **Type:** Passenger brake van (1887). **Owner:** Department of Education and Science. **Location:** National Rly Museum, York.
No: (DE 940281E). **Type:** Third brake (1888). **Owner:** Vintage Carriages Trust. **Location:** KWVR Haworth.
No: 1470. **Type:** Third brake (1895). **Owner:** Quainton Rly Soc. **Location:** Quainton Road.
Type: Passenger bk van. **Owner:** Peterboro Rly Soc. Ltd. **Location:** Wansford.

The brake third on the Keighley & Worth Valley Railway. R. Higgins.

Bass Ltd
Saloon 1889

The coach as preserved. Staffs C.C.

The Bass breweries at Burton-on-Trent were served by an extensive railway system with its own fleet of locomotives. During 1964 the railway network was abandoned as a result of reorganisation and greater use of road transport. The little saloon coach acquired by Staffordshire County Council in 1965 was an inspection saloon used to convey the brewery directors on tours of their empire. The photograph shows the varnished wooden body mounted on a wooden underframe with simple running gear. It is exhibited in company with a locomotive from the brewery.

Type: Saloon Coach (Four-wheel). **Owner:** Staffordshire CC Industrial Museum. **Location:** Shugborough Hall, Great Haywood.

City and South London
Pioneer Tube Stock 1890

The City & South London Railway was a notable pioneer: the first deep-level London tube railway and the first complete electric rail-was (as opposed to tramway) in the world, opened in December 1890. The trains consisted of small electric locomotives pulling three bogie cars. The railway gradually extended, until by 1907 it linked Euston with Clapham Common, and the rolling stock was increased to 165 coaches with 45 locomotives. The C&SL tunnels were smaller than the later London 'tubes' so that conversion, to enable through working, was inevitable. The original trains ran for the last time in November 1923 when the line was closed for reconstruction.

Coach No 10 is one of the earliest batch built by Ashbury Carriage & Iron Co in 1890. These coaches were dubbed 'padded cells', as no windows were provided apart from narrow glass ventilators under the roof line. The wooden body with flush-panelled sides is 26ft long, carried on two four-wheel bogies, with platforms at each end extended from the bogies. The platforms were for the use of the guards who worked the collapsible steel gates for entry from the station platform. Within the coach, the upholstery extends up the sides of the body — hence the nickname. The original exterior finish was varnished wood. This coach was originally on display at the old York Museum, but has now come south to Syon Park. In 1974 one of the bodies of the later all-steel coaches with clerestory roof and conventional bodyside windows was discovered near Hampton Court. This will eventually go on display at Syon Park.

No: 10. Type: 'Padded Cell' Coach (1890). **Owner:** London Transport Museum. **Location:** Syon Park.
Type: Clerestory Coach (1907). **Owner:** London Underground Railway Soc. **Location:** (in store).

'Padded Cell' No 30, as built. London Transport.

North Eastern Railway
Six-wheel Coach

Coach No 1111 in course of restoration. P. Brumby.

A standard range of six-wheel coaches was built by the NER from 1884. This coach, 32ft long and 8ft 6in wide, has two first-class, two third-class compartments and a luggage compartment. The solebars are wooden, sandwiched between steel plates — a form of construction common until the end of the century. No 1111 passed to departmental stock in 1924 and was converted to a tool van; the centre wheelset was later removed. The coach was purchased for preservation in 1973 and it is intended to restore it to NER livery, refit the interior and replace the missing wheelset.

No: 1111. **Type:** Luggage Composite. **Owner:** Members of NER Coach Group. **Location:** Grosmont.

Great Western Railway
Four-wheel Stock

1891

No less than 602 four-wheel coaches were built by the GWR between 1890 and 1902 for city surburban and branch line services. There were three or so different varieties, some built for use on GWR trains to the City over the Widened Lines in London. The GWR four-wheelers were distinguished by the way that the body ends sloped inwards at the bottom above the headstocks. The body ends themselves were panelled out with mouldings. The roofs of these coaches were of the three-centre profile — a flattened ellipse.

Of the four preserved, No 416 is a four-compartment brake third, 31ft in length and weighing 10½ tons. It was converted to a camping coach and numbered 9940 in the 1930s; subsequently it became a service vehicle. Two of the others came from three lots of four-wheelers — the last-but-one series to be built — which made up local sets in the Bristol area. No 290 (later 6290) is a first/second class composite, 26ft 10in in length and weighing 10½ tons. The two centre compartments were first class. No 975 is a five-compartment third, 28ft in length. For the record, they cost £525 and £458, respectively. Both became service vehicles, probably in the 1940s. No 8 (later 6008), of similar layout to No 290, was built for the Ruabon-Dolgelley service and finished its days as a mess van. The GWR four-wheelers were probably the last of this type in service on BR, as some were working as late as 1953.

No: 416. **Type:** Third brake (1891). **Owner:** Somerest Rly Museum. **Location:** (at GWS Didcot).

No: (6)008. **Type:** Composite (1900). **Owner:** Dart Valley Rly Assocation. **Location:** Buckfastleigh.

No: (6)290. **Type:** Composite (1902). **Owner:** Great Western Soc. **Location:** Didcot.

No: 975. **Type:** Third (1902). **Owner:** Great Western Soc. **Location:** Didcot.

A four-wheel third in service, 1935. H. C. Casserley.

Glyn Valley Railway
Closed Stock (2ft 3in gauge) 1892

The Glyn Valley Railway was a roadside tramway of 2ft 4½in gauge built to transport slate and granite from quarries to a canal. Originally horse-worked, it was operated by steam from c 1888 to

Coach No 14 as restored. Talyllyn Rly Co.

1933 (for passengers) and 1935 (freight). On closure, some of the coaches were sold and two are in service on the Talyllyn. These are very basic two-compartment box-like four-wheel coaches. No. 14- the Talyllyn number — was purchased in 1956 and put on new running gear, although the underframe and body structure are original. No 15 was originally a third, but now runs as a first; it was rescued from a farmyard in 1958. Upholstery, carpets and mirrors have been fitted in both coaches and a supplementary charge is made for riding in them. Nos 14/15 have been beautifully restored to the original GVR livery of medium green below the waist and cream surrounds to windows and doors. The body panels are elaborately decorated with gold-leaf and the GVR red circlet monogram is fixed to the centre of each side.

No: (14). **Type:** First (1892). **Owner:** Talyllyn Rly. **Location:** Towyn.
No: (15). **Type:** Third (1901). **Owner:** Talyllyn Rly. **Location:** Towyn.

Zillertalbahn (Austria)
Four-wheel Stock
1892-1901

The Welshpool & Llanfair's original passenger stock was scrapped by the GWR in 1936 so the preservation society has had to acquire new coaches. The most interesting of these came from the Austrian

760mm gauge Zillertalbahn in March 1968. The Zillertal coaches are 26ft 3in over the buffers, 8ft 1in wide and are four-wheelers. There is a verandah at each end and the open saloon seats 32. The bodies are wooden with vertical matchboard panelling and retain their original markings. Zillertal coach No 24 came from the SKLB when that line closed in 1957. It is of similar dimensions to the other coaches but has an elliptical roof overhanging the entrance balconies. The body-sides have four large droplight windows and the coach has a lavatory. It is good to record that the coaches made their trip from Jenbach, Austria, to Welshpool by rail and sea and not by road.

Nos: 14, 16, 17. **Type:** Zillertalbahn Second (1900/1). **Owner:** Welshpool & Llanfair Lt Rly. **Location:** Llanfair Caereinion.
No: ZB 24. **Type:** Salzkammergutlokalbahn Lavatory Second (1892). **Owner:** Welshpool & Llanfair Lt Rly. **Location:** Llanfair Caereinion.

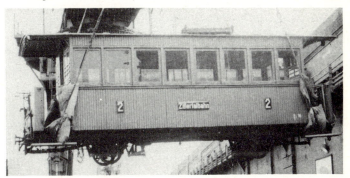

Zillertal No 16 arrives in Britain.

Liverpool Overhead Railway
Electric Motor Stock 1892

The Liverpool Overhead Railway was Britain's only elevated railway and operated from March 1893 until December 1956 — remaining independent throughout. It was electrified at 500/525V dc with side third-rail collection. The rolling stock was built by Brown, Marshall & Co Ltd with electric traction equipment from the Electric Construction Co Ltd and supplied in three batches between 1892 and 1895. The electric equipment was renewed in 1902 — to a higher performance — by Dick, Kerr & Co and again in 1919. Originally running in two-car sets, from 1914, all gradually were formed as

Original LOR stock in 1950. H. C. Casserley.

three-car units. In 1902/3 some original motor coaches were increased in width from 8ft 6in to 9ft 4in and trailers were converted or built new from 1914-36 to run with them. The first batch of motor cars were 45ft long, later vehicles 40ft only, and the trailers were 32ft in length (some 45ft). All coaches were of angular box-like appearance and the finish was varnished teak. From 1945-55 about 25 vehicles were rebuilt with aluminium-panelled bodies and sliding doors by the LOR.

On closure the scrap merchants despatched all the equipment except an original 1892 motor coach No 3, which was more or less in original condition, and two modernised coaches. No 3 was left intact and was later restored and is now on display in Liverpool. Of the rebuilt trailers, the body of No 7 was sold to scrap merchants and has since been rescued for the Southport museum project.

No: 3. **Type:** Motor Coach (1892). **Owner:** City of Liverpool Museum. **Location:** Liverpool.

No: 7. **Type:** Trailer Car (1895, rebuilt 1947). **Owner:** Southport Locomotive & Transport Museum Soc. **Location:** Southport.

Great Western Railway
Miscell Saloons 1894

These two fine coaches well deserve preservation. No 249, of 56ft body length and running on Dean 10ft wheelbase bogies, was originally built as a rather special family saloon. Only two months elapsed between the order being placed on the works and the coach entering traffic. Externally, it is remarkable by the fact of its 'royal' clerestory — with this type the ends of the clerestory slope down at the roof ends; nearly all the coaches with this roof were built for the 1897 Royal Train. Each end has observation windows and the roof itself extends to form a hood. Apart from the central kitchen and private compartment, the bodysides have large windows the length of the coach. One saloon was originally a dining room, the other a lounge. The external livery is now as the original — chocolate and cream with the panel mouldings picked out in black. No 249 was later a director's saloon, after 1907 numbered 9045. It was converted as District Engineer's saloon Plymouth some time between 1939-45 and numbered 80978. It remained on BR until 1965 (not used after 1963) and more recently has been attractively restored by the Dart Valley. I enjoyed a very directorial ride in it on the Torbay Steam Railway in 1974. To be recommended!

No 231 is a more conventional first-class family saloon with the standard late 1890s clerestory. The accommodation originally consisted of a saloon, compartment for the family, servants' compartment and brake compartment as well as a lavatory. The body is 45ft 6in in length. No 231 was constructed for £896. After 1907 it became No 9035. Conversion to an engineer's saloon came in 1941 when it was renumbered 80971 and allocated to the Newport Division, being withdrawn in 1967.

No: 249. Type: Director's Saloon (1894). **Owner:** Dart Valley Lt Rly. **Location:** Paignton.
No: 231. Type: First Brake Saloon (1896). **Owner:** Dart Valley Lt Rly. **Location:** Buckfastleigh.

No 249 as restored by the DVR. G.R. Hounsell.

Great Western Railway
Family Saloon
1894

This might be titled the preservation of 10, River Gardens, near Reading, for No 2511 served as a house from 1936 until 1972. No 2511 is a 31ft third-class family saloon with two saloons seating 34 and a central lavatory and washroom. The coach has a three-centre type semi-elliptical roof and weighed just over 12 tons when built. Such coaches would be hired by a family for a trip to the races, Henley Regatta, or for the summer holiday, and would be attached to any train. After the 1914-18 war this way of life changed and the saloons were relegated to service use or some even as vans for fruit traffic. No 2511 remained intact until 1936, when the body was sold as a riverside bungalow. In due course, a gabled roof was built overall and a verandah added to the surrounding structure. During 1969, Great Western Society officials were surveying other coach bodies on the riverside site and were invited to inspect what seemed to be a perfectly ordinary bungalow. Once inside they found that the interior was virtually intact, lovingly looked after and with the original coloured prints, labels, luggage racks and upholstery. Apart from a window fitted in one end there were few changes. In April 1972 the coach was vacated and the GWS set about demolishing the surrounding structure and subsequently removed the body to Didcot. Here it remains to be united with a suitable underframe. Remarkably enough the body is still completely rigid.

No: 2511. **Type:** Third Saloon (Six-wheel). **Owner:** Great Western Soc Ltd. **Location:** Didcot.

No photograph available.

North Wales Narrow Gauge Rly/Festiniog
Third-class Stock (1ft 11⅝in gauge)   1894

FR No 22 was built by Ashbury Railway Carriage and Iron Co Ltd. One of a pair, the other, No 21, was not restored after the 1955 reopening. The seven compartments are divided into two sections by a central partition but otherwise have shoulder-height divisions only inside. The coaches were too lightly constructed and gave poor service. The external appearance gives clue to the uninspiring nature of the type. In FR days they were seldom used in ordinary passenger service and relegated to quarrymmen's trains. No 22 was put on a steel underframe in 1967 — originally it was all wood — but more recently the body has been scheduled for replacement.

FR Nos 23 and 26 were built for the North Wales Narrow Gauge Railway by Ashbury, one of four or five. As built, they were part-glazed above the waist and of matchboard construction. In 1923, at the time of inter-working between the FR and Welsh Highland (ex NWNG Rly), the coaches were lowered in height and fitted with vacuum brakes. No. 23 was handed over to the FR at the closure of the Welsh Highland in 1937. No 26 is rather a mystery, since there was apparently no record of its existence in Welsh Highland days. It ended up on a farm, from where it was rescued to enter service on rebuilding in 1959. Subsequently they have both been put on to steel underframes and have been rebuilt with full-height glazed doors and new panelling.

Nos: 23, 26. Type: NWNG Third. Owner: Festiniog Rly. Location: Porthmadog.
No: 22. Type: FR Third. Owner: Festiniog Rly. Location: Porthmadog.

One of the Welsh Highland open thirds of 1894.

Cambrian Railways
Tri-composite Brake   1895

The Cambrian Railways operated a number of through coach workings over its lengthy main line to the Welsh coast. The bogie coach preserved at Didcot is a typical example of its type, one of a batch of 12. It is 48ft 3in long and 8ft 11in wide and was built by Metropolitan C&W. The bogies are of 6ft 6in wheelbase. The interior in its latter-day condition consisted of a small luggage compartment, two first-class, one second-class and two third-class compartments. In GWR days it was numbered 6277 and became a departmental coach, subsequently DW 80945. It was obtained from Oswestry in 1970.

At present the coach is in the condition in which it was purchased and is without seats or partitions and considerable work is needed to restore it. For this reason it has not been possible to provide a photograph.

No: 238. **Type:** Luggage Tri-composite. **Owner:** Great Western Soc Ltd. **Location:** Didcot.

No photograph available.

Padarn Railway
Slate Railways' Stock   1896

Slate from the Dinorwic Quarries was taken by a tramroad to Port Dinorwic form 1824. This line was replaced by a 4ft steam-operated railway in the 1840s. Road transport replaced the Padarn Railway in 1961. A quarrymen's service was run over the line until 1947, but none of the coaches used now remains.

The Director's Saloon. National Trust.

Twenty-three quarrymen's coaches were built in 1895 by the Gloucester RCW and the same builder supplied the attractive quarry directors' coach in the following year. This coach, with a balcony at each end, resembles a tramway vehicle but was equipped with swivel armchairs in the saloon, seating eight. Apart from its duties for the quarry officials it was also used to convey wages from the head office at the port to the quarries. It was employed in this capacity at least into the 1950s.

The original livery of the coach is in doubt; it is believed to have been painted yellow with the owners' coat of arms on the side panels. However, an early photograph shows it to have been in a darkish colour, at the time also possessing oil lamp pots; so at an early stage it may well have been painted red, the livery which it now carries.

Three small passenger wagons were used by officials on the quarry railway. These were all painted chrome yellow. They were also privileged to form trains for royal visitors on one or two occasions. The 'truck' at Penrhyn was the most ornate of the three. The other two are stored at the quarry workshops.

Type: Director's Saloon (4ft gauge). **Owner:** National Trust. **Location:** Penrhyn Castle, Bangor.
Type: The Yellow Coach (1ft 10¾in gauge). **Owner:** National Trust. **Location:** Penrhyn Castle, Bangor.

Lancashire, Derbyshire & East Coast Railways
Six-wheel Coach 1896

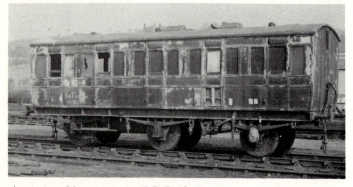

As purchased for preservation. J. B. Radford.

The Lancashire, Derbyshire & East Coast Railway was the last but one of the major railway projects to be built. Primarily for coal traffic, it operated a passenger service from Chesterfield to Lincoln, from Mansfield to Ollerton and Mansfield to Sheffield — the first starting in late 1896/early 1897. At first, ex-Great Eastern suburban coaches were used but orders for new stock were placed in 1896. These coaches were mostly six-wheeled and built by contractors; 64 old and new coaches passed to the Great Central, which absorbed the LDECR in 1907. The coaching stock livery was crimson red, lined out in yellow.

The third which has been preserved is probably the last LDECR coach to survive. It is an arc-roofed, five compartment third probably built by either Ashburys R C & Iron Co or Brown Marshall & Co. For its date of building it cannot be considered of very advanced design. Originally gas-lit, it was later converted to electric lighting. It was renumbered 1808 by the GCR — original number unknown — and was in passenger service until 1938, being converted to a staff sleeping van the next year. It was purchased in 1972 as BR No DE 950249. Extensive restoration is now being carried out by the Midland Railway Co Ltd on behalf of Derby Corporation.

No: GCR 1808. **Type:** Third (Six-wheel). **Owner:** Derby Corporation Museum. **Location:** Butterley.

Glasgow District Subway Co
Underground Stock (4ft gauge) 1896

This railway, Britain's other underground, was opened to traffic in 1897 as the Glasgow District Subway Company with an unusual 4ft track gauge. In 1935 cable traction was displaced by third-rail electrification at 570V dc. Following the formation of the Greater Glasgow PTE it was decided in 1974 to modernise the subway and replace the present rolling stock. The original cable cars were retained on electrification and rebuilt: the length of each car is 40ft 9in, the width 7ft 6in. The subway tunnels are 11ft diameter. The cars are similar internally to early electric trams, with longitudinal benches divided by two pairs of half-partitions. The entry platforms originally had gates, but from the 1940s sliding doors were fitted. The present rolling stock consists of 26 motor cars, 24 trailers and an electric locomotive. Two car trains are operated: each tube works traffic in one direction only and the trains are driven from the leading end only. Trailer No 41 was restored during 1974 at the subway's Govan works to the original company livery of cream and maroon from the present all-red paint scheme. This was done for a film documentary, but the coach has also re-entered service and will remain in use until 1977 when the new stock arrives. After then it will go to the Glasgow Transport Museum.

No: 41. **Type:** Trailer. **Owner:** Greater Glasgow Passenger Transport Executive. **Location:** Glasgow (Govan).

No 41 as restored and in service 1974. G Glasgow PTE.

North Eastern Railway
Inspection Saloon   1896

No 41 was built for the use of Vincent Raven, Assistant Chief
Mechanical Engineer of the NER, and used by him until 1910 when,
on becoming CME, he took over saloon No 1661 (page 21). This six-
wheeled vehicle is 30ft in length over the headstocks. There is an
open verandah at the 'trailing' end of the coach. The main saloon
seats eight and there is also a kitchen and an attendant's compartment. After 1910 No 41 was used by the NER's Chief Superintendent
and later by the railway's chief engineer. In LNER's days it became
No 241 and subsequently was fitted with roof lights for tunnel
inspection and stationed at York. The BR number was DE 900269.

No: 41. **Type:** Inspection Saloon. **Owner:** Keighley & Worth Valley RPS.
   Location: Haworth.

Great North of Scotland Railway
Six-wheel Stock   c1896

The two six-wheel coaches are of an unexceptional low-roof design —
the third has six compartments. Both coaches were purchased from
departmental stock — with the numbers as given above — in 1974.
At present they are used for storage but it is intended to restore
them to GN of SR condition in due course.

No: (DE 320010). **Type:** Third. **Owner:** Strathspey Rly. **Location:** Boat of
   Garten.
No: (DE 970204). **Type:** Full Brake. **Owner:** Strathspey Rly. **Location:** Boat of
   Garten.

West Coast Joint Stock
Saloon   1897

This vehicle is a curious hybrid, having started life as one of four
dining cars built at Wolverton Works for the West Coast Joint
Stock. It was withdrawn from BR service as an inspection saloon,
having acquired a new underframe and Gresley compound-bolster
bogies. This is the state in which it exists today, although painted in
Caledonian Railway livery. The reason for its present-day livery is
that it was transferred from WCJS stock to the Caledonian fleet and

Top: *NER Inspection Saloon in BR service.* K. Hoole.
Centre: *GN of S six-wheelers in LNER days.* LPC.
Bottom: *As restored.* T. Boustead.

numbered 41. In 1927, the body was mounted on a new 48ft underframe built by Birmingham RCW with four-wheel bogies — and it still carries the builder's plates today. As a dining car it had been a twelve-wheeled vehicle. Finally, in 1956, it assumed its present condition when rebuilt with an observation end similar to the front of a diesel railcar. At about this time it exchanged the LMS bogies for the Gresley 8ft wheelbase type. The non-observation end still retains the recessed doorway of its former existence. Although much of the bodyside panelling remains, some of the large saloon windows have modern sliding ventilators. The interior consists of an end observation saloon, a main open saloon, kitchen and guard's vestibule. As LMS and BR No 45018 it was in service until 1972, when purchased for preservation; it is now restored for main line running.

No: 41. Type: Saloon. **Owner:** Mr W. H. McAlpine/Flying Scotsman Enterprises. **Location:** Carnforth.

Great Northern Railway
Saloon No 706   1897

GNR No 706, a bogie clerestory saloon, was built as the railway's directors' saloon. In the general appearance and bodywork details it is a typical Doncaster product of the late 1890s exhibited to particular advantage following the excellent restoration to varnished teak finish by British Rail Engineering Ltd in 1971. The coach was LNER No 43909 and acquired Gresley double-bolster bogies. In BR days it was numbered DE 942090 and stationed at Hornsey.

No: 706. Type: Inspection Saloon. **Owner:** Bluebell RPS. **Location:** Sheffield Park.

As restored. J. Everitt.

Lynton and Barnstaple Railway
Bogie Coaches

No 2 as existing at Clannaborough Rectory. J. T. Palm.

Opened in May 1898 the Lynton & Barnstaple was closed by the SR in September 1935, having passed to the Southern (actually the L&SWR) at the Grouping. The L&BR had 17 coaches, all but one of which started the services in 1898. The 'odd man out', built in 1903, was little different. Apart from the 1903 coach which was assembled by the L&BR, the coaches were all built by the Bristol Wagon & Carriage Works. There were six different types: saloons, observation coaches and compartment stock. The design and construction of the vehicles was of a high standard and they were among the first in the country with rolling-bearing axleboxes — later removed. All but the 1903 coach were to the same dimensions: 35ft long and 6ft wide with wooden bodies on steel underframes. The original lighting was by oil lamps, later converted to acetylene. Steam heating was fitted by the SR to some coaches. The original L&BR livery was reddish-brown with white upper panels — later SR green.

After the closure of the line only three coaches escaped scrapping. No. 2 (SR 6992) has a brake compartment at one end and an observation platform at the other. It was sold in 1935 to Clannaborough Rectory, near Copplestone, Devon. When I saw it in 1961, apart from crude repainting similar to the L&BR livery, it was almost entirely intact, having been used as a summerhouse. No 15 (SR 6993) was a compartment coach (later actually a composite) and was sold for use as a hen house on a length of track at Snapper Halt, near the trackbed of the line. It remained there largely intact (even with its bogies) until 1959 when it was bought by the Festiniog. The FR extensively rebuilt it as a buffet car and it entered service in 1963 as their No 14. No 1 (similar to No 2) was also sent to Snapper in 1935, but its condition so deteriorated that by the early 1960s it was little more than a ruin.

No: 2. **Type:** Composite Saloon Brake. **Owner:** Private. **Location:** Clanna-borough (Devon).
No: 15. **Type:** Third Brake. **Owner:** Festiniog Rly. **Location:** Porthmadog.

Great Eastern Railway
Royal Saloon   1898

The Royal Family first patronised the GER (then the Eastern Counties Railway) in 1847 and has continued to do so since. In GER days the royal trains to Wolferton (for Sandringham) started from St Pancras. The GER built royal saloons in 1864, 1894 (six-wheel), 1897 (bogie), and the saloon now preserved in 1898. Finally, in 1901, came another vehicle known as the Queen's Saloon. After 1923 the former GN/ECJS royal train displaced the GER vehicles.

No 5 was built for the use of the Princess of Wales (later Queen Alexandra) and closely followed the introduction of the GER's first bogie main-line stock in 1897. The coach is 48ft 3in over the ends and 8ft 6in wide, having an arc roof. Fox's standard pressed-steel 8ft wheelbase bogies were originally fitted. The teak-built body is similar to the standard GER main-line stock and it had a steel underframe (later timber). As built, the interior consisted of a smallish (12ft) saloon at each end. The centre of the body had a side corridor from which led a smoking compartment, servants' compartment and lavatory. Interior panelling was in walnut and satinwood. Gas lighting was fitted and the coach had dual brakes. The exterior finish was varnished teak lined out in gold.

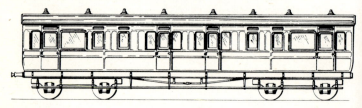

No 5 as built.

In 1925 No 5 was converted for the use of the civil engineer and renumbered 65. A number of interior alterations were made and a pantry installed. Observation windows were fitted in the body ends. In 1944 No 65 was again altered; the ordinary compartment was re-shaped for guard's use, a handbrake was fitted, Gresley 8ft bogies replaced the original type and soon afterwards the coach was renumbered DE 960900. After 1948 it was transferred to Sheffield and continued as the civil engineer's saloon until 1972, when it became a mobile office. Most of the interior partitions and panelling remain. It is intended to restore the coach as much as is practicable.

Another GER saloon has been acquired by the same owners. This is a six-wheel engineer's saloon originally No 14 built in 1889. It was latterly a mess van numbered DE 960903 and located at Doncaster.

No: 5. **Type:** Royal Saloon. **Owner:** Mr W. H. McAlpine/Flying Scotsman Enterprises. **Location:** Carnforth.

Corris Railway
Saloon Coach (2ft 3in gauge) 1898

A line-up of the Corris Railway bogie saloons. LPC.

The Corris was worked as a steam line from 1879, a passenger service starting in 1883. It later became part of the GWR and closed in 1948 although the passenger service finished in 1931. From 1898 there were eight bogie saloon coaches and two bodies were sold on withdrawal of the passenger services. The coach now on the Talyllyn was built by the Metropolitan C&W Co Ltd in 1898. It is a tramway-type vehicle with a central entrance vestibule and large glazed saloon sides and ends. It was recovered from a garden in

Gobowen (having been sold by the GWR at Oswestry), but on inspection at Towyn it was apparent that it needed complete reconstruction. A new underframe and bogies were required and it entered service in 1961. The coach has been excellently finished in the original Corris livery of chocolate brown with yellow lining.

More recently the Corris Railway Society have opened a museum at Corris where they have the skeleton of one of the bogie coaches. A replica of an 1895 bogie coach is being built for display in about three years.

No: (17). **Type:** Third. **Owner:** Talyllyn Rly. **Location:** Towyn.

Great Northern Railway
Non-corridor Stock   1898

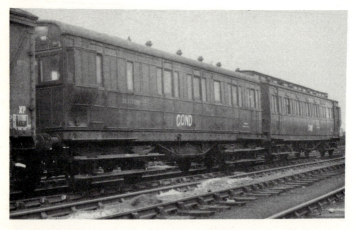

The coach as purchased for preservation. G. J. Holt.

As already noted, the GNR persisted with non-bogie stock into the 1890s. In 1893 six sets of corridor dining car trains were built for the East Coast Joint Stock by outside contractors but of the customary Doncaster pattern of six wheels, squared mouldings and windows. Only the dining cars were bogie stock: these had semi-ellipse profile low roofs. From 1896 the GNR built similar low-roof bogie stock for its own services: 45ft in length mounted on 8ft wheelbase bogies. These survived into the late 1940s/early 1950s. No 2856 is a lavatory brake non-corridor composite of this type built at Doncaster with four compartments, the lavatories being between the two sets of compartments. The teak-bodied coach has characteristic guard's lookouts. On withdrawal in October 1950 it was converted as a 'stator staff car' for the use of C A Parsons Ltd's personnel accompanying 'out of gauge' loads of electrical equipment. As such, some alterations were made: windows were cut in the brake compartment end. It was numbered BR No DE 320051 — previously LNER No 42856. Purchased in 1971 it is undergoing restoration.

No: 2856. **Type:** Lavatory Brake Composite. **Owner:** Newcastle Coach Group.
  Location: Grosmont.

Great Western Railway
Bogie Brake Van No 933 1898

Clerestory roof coaches were expensive to build and maintain so that some railways, particularly the GWR, built local service and non-passenger stock with three-centre or semi-ellipse roofs. During the 1890s the GWR produced a standard 40ft length design as passenger train brake vans and parcels/luggage vans. These were of 8ft width, running generally on Dean 8ft 6in (some 6ft 4in) wheelbase bogies. The last appeared in 1906. No 933 has four sets of double doors to each side and still retains its guard's lookouts. Vehicles of this type disappeared by the early 1950s, but No 933 has survived in particularly good condition, as for some twenty years, until purchase in 1968, it was kept under cover as a service vehicle at Hockley Goods Depot, Birmingham.

No: 933. **Type:** Passenger Brake Van. **Owner:** Private. **Location:** Didcot.

A similar vehicle but without guard's lookout.

East Coast Joint Stock
Luggage Third   1898

Within the period 1888-1906 the East Coast Joint Stock progressed from the prevailing six-wheel type (page 40) to modern elliptical-roofed bogie stock designed by H. N. Gresley. In the middle of this period, the late 1890s, there were two principal clerestory-roofed ECJS types: the massive twelve-wheel 59-62ft stock designed by Doncaster, and the neater eight-wheel 55ft coaches — such as No 12 — which were the work of David Bain, the NER carriage superin-tendent, and built at York. The Bain design was the more attractive of the two with wide vestibules and glazed lights in the bowed body ends. The GN Doncaster-built coaches, although imposing, had cramped compartments and were expensive to maintain; with-drawals started in the 1920s. Both types had teak-panelled bodies with the usual square-cornered body mouldings. The clerestory sloped down at the roof ends.

No 12 is a 55ft body coach, 9ft in width and with 8ft wheelbase bogies, of the type introduced for ECJS stock in 1893. It is gas-lit and has an unusual layout of entrance vestibules at each end, corridor-side doors and an external door to each compartment. It has six compartments and a luggage compartment. The glazed body end lights have not been restored. It was withdrawn in 1952 as No 41805, having been previously transferred from the East Coast stock

A York-built ECJS third — of similar pattern to No 12 — in service in 1955.
E. Oldham.

to LNER (GN) allocation. No 12 was then excellently restored to ECJS teak livery for the centenary celebrations of the GN main line in October 1952 and exhibited at Kings Cross. Following this it has never since been on permanent public display and is stored at present.

No: 12. **Type:** Luggage Third. **Owner:** Department of Education and Science. **Location:** (in store).

Great North of Scotland Railway
Royal Saloon   1898

In BR service in 1961. M. Pope.

This coach was built as first-class saloon No 1 at the GN of S Inverurie Works for private hire duties. The GN of S line from Aberdeen to Ballater was convenient for Balmoral and so at the beginning of Edward VII's reign in 1901 No 1 was used as a royal coach. The railway made up a royal train for the monarch which was used in 1903/4 for Edward VII's trips to the St Leger race at Doncaster. Saloon No 1 was also patronised on other occasions by the King until his death in 1910. Thereafter it was employed once again for private hire, but the LNER transferred it to departmental stock as a saloon in 1924. It was withdrawn for scrap by BR in 1965, but fortunately rescued at the eleventh hour by the SRPS. The BR running number was 982002.

The coach, with a lowish clerestory to the roof, is 48ft long and 8ft 6in wide. The layout originally consisted of two saloons, one first-class compartment, an attendant's coupé at one end and pantry facilities. The middle saloon could be converted to sleeping accommodation.

The SRPS has used the coach, as yet unrestored externally, on rail-tours, but its lack of corridor connections is a disadvantage, although it may be used on private hire.

No: 1. **Type:** Saloon. **Owner:** Scottish RPS. **Location:** Falkirk.

Metropolitan Railway
'Bogie' Stock   1898

To provide a higher standard of comfort for its services to Chesham, Amersham and Aylesbury and beyond than the 'Jubilee' stock and rigid eight-wheelers, the Metropolitan Rly put 54 coaches of 'Bogie' stock into service between 1898 and 1900. The earliest were built by Ashbury RCW, others by Cravens and the Met works at Neasden. Six-coach trains were made up from this stock. The 39ft 6in length coaches were conventional teak-bodied arc-roofed vehicles, 8ft 3in wide and running on 7ft wheelbase bogies. Electric lighting was provided from the start. These were the first bogie eight-wheeled coaches on the Met. Between 1906 and 1924 all the 'Bogie' stock was converted to form electric multiple-unit stock working to Uxbridge and, after 1932, to Stanmore. The livery was varnished teak with cream panels at waist and cantrail. In London Transport days they were teak overall. New multiple-unit stock replaced these trains from 1938 but most was held in reserve until 1945/6. Meanwhile six coaches were converted for the steam-worked push-pull service between Chalfont & Latimer and Chesham in 1940/1, to form two three-coach trains. Each train consisted of a former four-

The Bogie stock on the Bluebell, 1961. S. C. Nash.

66

compartment motor coach with louvred panels to the switchgear compartment (No 394 as preserved); a former composite trailer, downgraded to third (Nos 368/412 as preserved); and a driving trailer third (Nos 387/400). The Chesham branch was electrified in September 1960. Four of the coaches were bought by the Bluebell and one was kept for the British Railways collection. The coaches on the Bluebell were repainted brown and cream and bore the brunt of that railway's services from 1961, returning to LT for the Underground centenary celebrations in May 1963. However, their condition deteriorated and they are not currently in use on the Bluebell, although No 387 is in the course of thorough renovation.

The milk van was built in 1896 for milk churn traffic from the Chilterns to London. It was in service use from 1936-63, when it was restored to Metropolitan Railway varnished teak finish for the centenary celebrations; later it went on display at Clapham.

No: 368(515). **Type:** Composite (1900). **Owner:** Bluebell RPS. **Location:** Sheffield Park.

No: 387 (518). **Type:** Driving Trailer Third (1898). **Owner:** Bluebell RPS. **Location:** Sheffield Park.

No: 394 (512). **Type:** Third Brake (Motor Car) (1898). **Owner:** Bluebell RPS. **Location:** Sheffield Park.

No: 400 (519). **Type:** Driving Trailer Third (1900). **Owner:** Department of Education & Science (in store).

No: 412 (516). **Type:** Composite (1900). **Owner:** Bluebell RPS. **Location:** Sheffield Park.

No: 3. **Type:** Milk Van. **Owner:** London Transport Museum. **Location:** Syon Park.

Duke of Sutherland's Saloon 1899

The Duke of Sutherland, whose ducal seat was at Dunrobin Castle some 90 miles north of Inverness, was the principal landowner in the county of Sutherland. The third Duke was instrumental in promoting the completion of the railway between Inverness and Wick, part of the route being his private railway, merged into the Highland Railway in 1884. Subsequently the Duke retained the right to run special trains from Inverness to Wick, a right which was exercised at least as late as 1938.

Apart from owning a locomotive and small saloon, the Duke had a special saloon for long-distance travel. This vehicle was built at the LNWR's Wolverton works to the design of C. A. Park, the carriage superintendent. In its general appearance it can be considered the

The Saloon in pre-1914 days. Inverness Public Library and Museum.

precursor of the LNWR Royal Train vehicles. The Duke's saloon is 57ft over the balconies and 8ft 6in wide, with a steel underframe, and runs on 8ft wheelbase bogies. The clerestory roof body has balconies at each end. The interior has an attendant's compartment at one end, then a dining saloon, and from there a side corridor provides access to two sleeping compartments. At the other end the day saloon seats four. The interior is panelled in lincrusta with most furnishings in green. The exterior is painted dark green for the lower panels with cream upper panels lined out in gold. Electric lighting was provided from the start. Cooking was by oil and a stove supplied hot water heating. In 1949 the Duke's locomotive and saloons were sold. The saloon described was on display at Clapham.

No: 57A. **Type:** Duke of Sutherland's Saloon. **Owner:** Department of Education & Science. **Location:** National Rly Museum, York.

London & North Western Railway
Royal Train Saloons   1900-3

By the end of the nineteenth century rolling stock design had come similarly to the end of an era. Within ten years the elliptical roof-corridor coach with relatively simple furnishings and electric lighting would become standard on most of the largest railways. Just at the turn of the century, the pinnacle of late Victorian design was reached with the massive twelve-wheeled clerestory stock built for the West and East Coast Joint Stock fleets.

The prototype for the LNWR/West Coast Joint Stock twelve-wheelers was probably the Duke of Sutherland's Saloon (page 67). The LNWR/WCJS Royal Train vehicles made up one of the finest set trains to have run at any time: the coaches had an impressive outline, the twelve-wheelers a grace all of their own, while the livery provided the final flourish. The complete train was usually of 11 coaches: a brake first couchette (for staff) at each end, two dining cars (described below), a sleeping car, the King and Queen's saloons and six 57ft semi-royal coaches. Four coaches from the train have been preserved and the two brake firsts (now numbered M 5154M/M5155M) built in 1906 are still allocated to the Royal Train in 1975.

The two dining saloons, Nos 76/77, were of the magnificent 65ft 6in length twelve-wheeled clerestory-roofed design evolved by the LNWR's carriage superintendent, C A Park. Catering and sleeping vehicles were built to this specification during the 1900s and continued to be constructed to the same general outline, although without the clerestory roof, into the early days of the LMS. No 76 is 65ft 6in over the vestibules, the width over the mouldings is 8ft 6in and the height from rail to the top of the clerestory is 12ft 7½in. The body is of wood carried on a steel underframe with 11ft 6in wheelbase six-wheel bogies. The photograph shows the principal features of the design: a fairly shallow clerestory with decklights sloping down at the roofends, a massive body with nicely detailed mouldings, round-topped windows set in pairs surmounted by ventilators and a recessed entrance to the vestibule protected by the overhanging roof eaves. Electric lighting is installed and the kitchens have gas cookers. As built there were fixed seats for 20 passengers: in the present form there are loose chairs around a central table. The kitchen and pantry are at one end and a lavatory and WC at the other.

King and Queen's Saloons.

69

Dining Car No 77. C. Scott Shaw.

No 76 was exhibited at the Paris Exhibition of 1900, where it gained the Grand Prix. It was included in the LNWR royal train formed in 1903 and continued to run in the LMS/BR royal set until replaced by a new car in 1956. It was restored at Wolverton and subsequently exhibited at Clapham. No 77 is a standard LNWR kitchen dining car to the same overall dimensions and built in 1901. One-third of the interior is made up of the kitchen and pantry. The rest, includes two dining saloons, seating eight and twelve. From 1901-5 it was in general service and was then transferred to the LNWR Royal Train for use by the staff. It was included in the Royal Train until 1966, when it was replaced by a BR standard first-class diner. It has not yet been restored externally.

The set pieces of the LNWR Royal Train were the King and Queen's saloons built at Wolverton works in 1903/4, which replaced Queen Victoria's 1869 saloon (page 19). These are to the same general specification as the dining cars already described. However, the entrance vestibules are greatly enlarged, fitted with double doors and recessed into the bodysides. Folding steps are fitted. Each vehicle contains a day saloon and night saloon, with appropriate regal chairs, sofas and silver-plated bedsteads. During the 1914-8 war, when the Royal Train was used by tours for the monarch, silver-plated baths were installed in the dressing rooms of both cars.

The exterior finish of both cars was magnificent. They were painted in the standard purple-brown and white LNWR livery, but the bodysides were adorned by hand-painted armorial bearings, golden lions' heads on the headstocks and gold-plated door and grab handles. At the wish of King George V the LNWR livery was retained for the whole train until the 1939-45 war, when it was painted LMS red. Two new royal saloons were built in 1941: these and two other 1942-built saloons eventually replaced the 1903 cars, both of which were on display at Clapham.

No: 76. **Type:** Royal Dining Saloon (1900). **Owner:** Department of Education & Science. **Location:** National Rly Museum, York.

No: 77. **Type:** First Kitchen/Dining Car (1901). **Owner:** Quainton Rly Soc Ltd. **Location:** Quainton Road.

No: (800). **Type:** King Edward's Saloon (1903). **Owner:** Department of Education & Science. **Location:** National Rly Museum, York.

No: (801). **Type:** Queen Alexandra's Saloon (1903). **Owner:** Department of Education & Science. **Location:** National Rly Museum, York.

North Eastern Railway
Clerestory Stock   1900

No 3453 makes an interesting comparison with the GWR non-corridor clerestories already described. Generally speaking it can be said that the GWR, NER and MR built the best clerestory stock of the period. The NER coaches were to the designs of David Bain, carriage superintendent who went to the Midland in 1902. In comparison with the GWR coaches the NER clerestory was of neater design. A large number of bogie clerestory vehicles were built by the NER from 1895 to 1906.

No 3453 was built as a standard eight-compartment third, 52ft in length. The most noticeable feature is the pair of portholes cut in the

No 3453 as purchased for preservation. P. Brumby.

body end. From 1905 the NER introduced push-pull trains — which the railway termed 'steam autocars' — first brought into service between Hartlepool and West Hartlepool. No 3453 was converted to a driving trailer in 1906, hence the windows. Two compartments were converted to form a driving cab and luggage compartment, and the adjoining compartment became first class. In 1921 the control equipment was removed and the coach remained a brake composite until 1952. It then became a mobile office and was purchased for preservation in 1970. Very extensive restoration has been carried out at Hull. It is due to go to the North York Moors Railway.

Nos 818 and 1149 are standard 52ft clerestory luggage composites with seven compartments, the three centre ones being first class. They became all-thirds around 1930. Originally air-braked and gas-lit, they were converted by the LNER to vacuum braking and electric lighting. Both were sold about 1949 to the National Coal Board for workmen's trains in Northumberland. A third coach, No 3071, was bought by Beamish but has been dismantled after fire damage. Although sound, Nos 818/1149 require considerable renovation.

No: 3453. **Type:** Third Auto-car (1904). **Owner:** North Eastern Rly Coach Group (Hull).

No: 818. **Type:** Luggage Composite (1900-5?). **Owner:** North of England Open Air Museum. **Location:** Beamish.

No: 1149. **Type:** Luggage Composite (1900). **Owner:** North of England Open Air Museum. **Location:** Beamish.

South Eastern & Chatham Railway
Family Saloon   1900

This coach was built at Ashford soon after the formation of the SE & CR and exhibits the characteristic vertical body mouldings of that railway. It was one of a pair of non-corridor family saloons. The body is 38ft long and 8ft wide. As originally built the layout consisted of a central transverse corridor leading to the large saloon, which seated 13 on couches and in chairs. Beyond the saloon was a luggage compartment. On the other side of the corridor were lavatories and an ordinary compartment for use by the servants. The interior was panelled in ash and olive wood. It was one of the first SECR coaches to have electric lighting. In 1908 it was converted to an invalid saloon, the corridor and saloon were enlarged into one area and double doors fitted for the entry of stretchers. It became Southern Railway No 7913 and was sold to the Longmoor Military

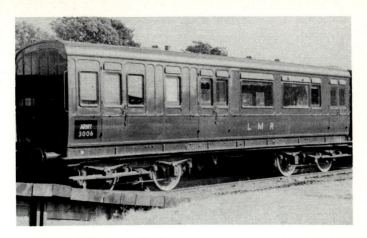

The coach in LMR days. J. Gardner.

Railway in 1936. Here it was used as an officers' saloon and apart from the removal of the continuous stepboards remained complete. It was used on the LMR until 1970, latterly Army No 3006. It is intended to restore it to SECR livery in due course.

No: 177. **Type:** Family Saloon. **Owner:** Transport Trust. **Location:** Bewdley.

London & South Western Railway
Bogie Stock   1900

In the mid-and late-1890s the L&SWR re-equipped much of its coaching stock with a fleet of modern bogie coaches characterised by their design of semi-elliptical roofs. The majority of these were non-corridor and a number were composites or tri-composites to cater for the various through coach workings to the West Country. Some had wooden underframes and bodies, although steel was becoming more common for frames by the early 1900s.

The most notable of the preserved L&SWR coaches is the tri-composite brake, LSW No 3598, later SR 6474. The layout of this coach is tidy: two first-class, two second-class and two third-class compartments — each with access to lavatories. No 3598 was carefully restored in London & South Western salmon and brown livery in 1948 on the occasion of the centenary of Waterloo station.

73

It has not subsequently been on public display. No 320 was originally an all-wooden 48ft third, No 1228, built in 1900. It was rebuilt in 1935, when it was mounted on a new 58ft underframe; the body was cut in half and a section including the two lavatories inserted. No 320 seats 88 and was withdrawn in 1959. It is currently being restored. No 5065, a six-compartment composite, was converted to a camping coach (No 31) in 1953/4 and originally allocated to the Amberley (Sussex) site on the SR. DS 1119, original number not known, was a three-compartment brake third subsequently converted for engineer's use and now serves as a workshop.

No: 3598. **Type:** Tri-composite Brake (1903). **Owner:** Department of Education & Science (in store).

No: SR 320. **Type:** Third (rebuilt 1935). **Owner:** Bluebell RPS. **Location:** Sheffield Park.

No: SR 5065. **Type:** Lavatory Composite (c1900-5). **Owner:** Tenterden Rly Co Ltd. **Location:** Rolvenden.

No: (DS 1119). **Type:** Third Brake (c1900-5). **Owner:** Merchant Navy Loco Pres Soc. **Location:** Ashford.

No 320 (leading coach) arrives on the Bluebell in 1960. D. Cross.

Great Western Railway
Dynamometer Car   1901

A dynamometer car is used for the road testing of locomotives, to measure the effectiveness of the locomotive as traction power. The

The car at the end of its days with BR. R. C. Riley.

GWR car is 45ft long over the end panels with the roof extended at each end to form overhanging eaves. It has a Royal clerestory (see page 48). The bodyside bulges amidships for observation purposes and end windows are fitted. The interior consists of two saloons, a lavatory and instrumentation including a Hallade recorder with a flangeless retractable wheel for recording speeds. It was originally numbered 790, later No 7, and painted in standard GWR livery. After the 1939-45 war No 7 was extensively used for the controlled road testing of a variety of locomotives in which all the variables affecting power output were measured under monitored conditions. It was used during the famous locomotive exchanges of 1948 and continued hard at work during the 1950s. Some time after 1954 sliding ventilators were fitted to some of the large windows — probably the more noticeable change in 50 years. Swindon produced a new dynamometer car in 1961 and the old vehicle was retired. Fortunately, despite BR's decision not to preserve No 7, it was rescued by the Dart Valley as a fitting tribute to Swindon's expertise in testing steam locomotives.

No: 7. **Type:** Dynamometer Car. **Owner:** Dart Valley Lt Rly. **Location:** Buckfastleigh.

Great Western Railway
46ft Clerestory Stock   1901

These two coaches are typical examples of GWR clerestory non-corridor stock, both eight-compartment thirds. They are 46ft 6in

A rake of Dean non-corridors including two thirds 1951. R. C. Riley.

long and the only major difference between them is that No 1941 is 8ft in width and No 1357 six inches wider. Both were fitted from the start with steam heating, and lighting was by gas. As was standard practice after 1895 for vehicles under 50ft in length, they are mounted on Dean 8ft 6in wheelbase bogies. One detail of interest is the blank panel in the centre of each end of both coaches: this was apparently intended to facilitate the fitting of gangway connections at a later date. Well over 200 of these thirds were built and they were long lived: the last were not officially withdrawn until 1958 (probably appropriations for temporary service use), but some were certainly in use on workmen's trains until about 1953. No 1941, latterly in service stock, has much of the interior remaining including the gaslight globes. Restoration is proceeding well and it promises to be a very nice example of preservation.

No: 1941. **Type:** Third (1901). **Owner:** Great Western Soc Ltd. **Location:** Didcot.
No: 1357. **Type:** Third (1903). **Owner:** Great Western Soc Ltd. **Location:** Didcot.

North Eastern Railway
Inspection Saloon   1903

This is a particularly fine clerestory saloon built in June 1903 for the use of the NER traffic officers. The bow-ended body, 41ft 6¾in over the ends, is built on a wooden underframe. Fox's pressed steel

8ft wheelbase bogies are fitted. The prestige effect of No 305 is enhanced by an unusual feature of projecting lookouts on each side of the body, both 12ft 6½in in length. These are alongside the main saloon. At the other end is a smaller saloon and both ends have observation windows. The coach has the customary central kitchen and toilet. No 305 has remained in almost its original condition and the interior particularly has kept its fine wood panelling, brasswork fittings and furniture. This is complemented by the very high standard of restoration to the original crimson lake livery and full lining out which was carried out at the York works of British Rail Engineering Ltd in 1973.

No 305, as far as is known, spent the whole of its life at York. After 1923 it was the LNER's Northern Area general superintendent's saloon and numbered 2305. In BR days it was used by the chief operating officer of the North Eastern Region/Eastern Region as No DE 902177.

No: 305. **Type:** Inspection Saloon. **Owner:** Mr W. H. McAlpine/Flying Scotsman Enterprises. **Location:** Carnforth.

As restored in 1973. R. Wildsmith.

Dundalk, Newry & Greenore Railway
Six-wheel Coach — (6ft 3in gauge)

No 1 arrives at Belfast Museum. Belfast Folk & Transport Museum.

The Dundalk, Newry and Greenore Railway opened in 1876 was owned by the LNWR as part of its development of Anglo-Irish traffic — principally freight in this case. The two lines comprising the DN&GR were worked by LNWR-built locomotives and rolling stock. After 1923, although passing to LMS control, the railway continued to maintain its pre-Grouping flavour and livery. In 1933 the Great Northern Railway (Ireland) took over the working and maintenance of the DN&GR, but it continued in British ownership until closed by the British Transport Commission on 31 December 1951.

No 1, like the other coaches, remained in the LNWR livery of purple-brown and off-white until closure of the railway. It was built at Wolverton to replace the original passenger stock. It is of typical LNWR design, with 2½ first-class and 2 second-class compartments; the end half-first-class compartment is a coupé with end windows, apparently to aid the guard in shunting manoeuvres. The coach is 34ft 6in long.

No: 1. **Type:** Composite Coupé. **Owner:** Belfast Folk & Transport Museum.
Location: Belfast.

Midland Railway
Clerestory Stock

Thomas Clayton, the Midland's carriage superintendent, introduced a new design of clerestory coach in 1897. These were 60ft long and had rectangular body mouldings, square-cornered windows with toplights above, and steam heating. Except for the dining cars, the coaches were non-corridors. The roofs were semi-elliptical, the clerestory appearing more obviously part of the roof than with some other designs. No 2944, LMS No 7263, a side-corridor coach, has 2½ first-class and 2½ third-class compartments. It was converted to a tool van in 1947 and was later a bridge-testing unit as DM 198829. As the photograph shows, it has retained most of its external fittings. The third brake, also a side-corridor vehicle, was extensively rebuilt as a tool van and was formerly at Bath Green Park engine shed. It is complete with all the tools and will remain in converted state.

In 1902 David Bain came to the Midland from the North Eastern to succeed Clayton. The general outline of his coaches was similar to those of his predecessor, although rounded body mouldings and windows reappeared, with deeper panels under the cantrails. No 634, LMS No 3186, was converted for civil engineer's use in 1951. It was bought by the Keighley & Worth Valley RPS in 1965 and later sold to Derby Corporation. Unfortunately the body is beyond repair so that in due course a body from another coach will be fitted on to the underframe and bogies of No 634. The third, at Haverthwaite, was converted to a riding and mess van.

No: 2944. **Type:** Composite Brake (1904). **Owner:** Derby Corporation. **Location:** Butterley.

No: (DM 198715). **Type:** Third Brake (1905). **Owner:** Derby Corporation. **Location:** Butterley.

No: 634 (see below). **Type:** Composite (1910). **Owner:** Derby Corporation. **Location:** Butterley.

No: (DM 395031). **Type:** Third (1913). **Owner:** The Lakeside Rly Soc. **Location:** Haverthwaite.

No 2944 as purchased for preservation. J.B. Radford.

Great Western Railway
Dreadnought

We come now to a milestone in passenger coach design: the introduction of elliptical roof stock. The GWR 'Dreadnoughts' — so called after the contemporary all-big-gun battleships — were even more notable as probably the largest general service coaches to run in Britain until the BR Mark III stock of the 1970s. The big 'clipper' bodies, 68-70ft long, bulged outwards to a 9ft 6in width to make full use of the GWR's broad-gauge loading gauge. There was enough opposition within the GWR to coaches of this size, but criticism also resulted from the unusual layout of the interiors. The side-corridor stock had no exterior doors to the compartment side, so that entry and exit were by way of the large end and central vestibules. In addition, the internal corridors crossed from one side to the other, divided by the central vestibule. Most were electrically lit from the start but some, such as No 3299, a nine-compartment third built by Birmingham RC&W, were originally gas lit.

The 'Dreadnoughts' first went into service in May 1904 and a variety of coaches including sleeping cars were built up to 1907. However, the general 70ft corridor stock built from 1906 included few of the more unusual features of the 'Dreadnoughts', which in practice proved unpopular with operating staffs. Withdrawals began in 1949, but some survived until 1956. No 3299 was used as sleeping accommodation for restaurant car crews at Newquay from 1954. Fortunately, David Rouse stepped in to preserve it and it is good to know that No 3299 is now being restored to first-class condition.

No: 3299. **Type:** Third. **Owner:** Great Western Soc Ltd. **Location:** Didcot.

'Dreadnought' third No 3277 as built.

Brake third No 4168 as restored 1969. Alastair McIntyre.

On taking over the Isle of Wight railways the SR gradually sent over new stock from the mainland, so that by nationalisation the typical IOW train consisted of ex-LBSC or SECR bogie stock hauled by ex-LSWR 0-4-4Ts. Westinghouse air-braking was a particular feature of the Island's railways. The first bogie stock in quantity came to the IOW in 1931 and ex-LBSC vehicles were transferred from 1935. The coaches sent over to the Island were altered in varying degrees and flush steel panelling applied. Steam operation with the old rolling stock ceased on 31 December 1966, but hard work locally has seen the preservation of an 0-4-4T and matching coaches (see also page 108).

One of each type of the ex-LBSC coaches used on the Island has been preserved — out of a total of 52. No 4168 is a five-compartment brake third; some of these survived on the mainland in push-pull sets until the early 1960s. Overall length and width are 57ft 7in and 8ft 10in respectively. No 2416 is a nine-compartment third. No 6349 is an eight-compartment composite with side internal corridor, also 57ft 7in by 8ft 10in. A similar vehicle, No 6237, from a mainland push-pull unit, was reserved by the Bluebell but mistakenly sent for scrap by BR. The LBSC coaches are distinguished by the shallow arc profile to the roof. Incidentally, the

81

body of an ex-LCDR brake third, No 4115, has recently been acquired by the Wight Locomotive Society.

No: 4168. **Type:** Third Brake (1905). **Owner:** Wight Locomotive Soc. **Location:** Haven St (IOW).

No: 2416. **Type:** Third (1916). **Owner:** Wight Locomotive Soc. **Location:** Haven St (IOW).

No: 6349. **Type:** Composite (1924). **Owner:** Wight Locomotive Soc. **Location:** Haven St (IOW).

North British Railway
Third   1905

This coach is an arc-roofed non-gangwayed lavatory third built at the NBR's Cowlairs Works. There are seven compartments and it is 57ft long. Later LNER No 31748, it was converted in 1947 to a mess and dormitory van and as a result was rebuilt externally with full-length planking covering the lower body panels. At present it is used for staff purposes at Boat of Garten, having come from Inverness as DE 970012 in early 1973.

No: 1748. **Type:** Lavatory Third. **Owner:** Strathspey Rly. **Location:** Boat of Garten.

As preserved. Bill Roberton.

North Eastern Railway
Dynamometer Car

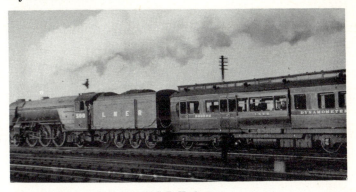

The car in early post-war days. C. C. B. Herbert.

In 1903 the NER borrowed the GWR's dynamometer car to test one of its newly constructed V class 4-4-2 locomotives. As a result, the NER decided to build its own dynamometer car. In general appearance the NER car is very like the Swindon product and the drawings of equipment in the GWR car were lent to the North Eastern design staff. The car, No 3591, was built on a standard 52ft underframe in 1906. The bodywork followed the NER clerestory outline, but with a projecting observation lookout on each side of the recording saloon. The comparative tests with a NB 4-4-2 between Newcastle and Edinburgh in 1907 were probably the first major task for the car, which was based at Darlington.

At the Grouping, the NER dynamometer car was the only one inherited by the LNER and was numbered 23591. In due course it received Gresley double-bolster bogies to improve its riding. Sir Nigel Gresley was keen that the 'Big Four' companies should jointly construct a locomotive testing plant from the late 1920s, but only in 1937 did the LNER decide to build a new dynamometer car to replace the 1906 vehicle. Construction of the new car was authorised in 1937 but it was destroyed by fire during construction in 1940. Meanwhile the old NER car had been involved in its most famous exploit, as it was used for the record-breaking run using A4 Pacific *Mallard* in 1938 when 126 mph was attained. The car, by then numbered 902502, was used during the 1948 locomotive exchange trials and it was not until 1951 that the replacement car appeared.

Subsequently the NER car was selected for preservation by the then BTC and it was exhibited at Clapham, restored to its 1947 condition and coupled to *Mallard*.

No: (3591) LNE 902502. **Type:** Dynamometer Car. **Owner:** Department of Education & Science. **Location:** National Rly Museum, York.

Great Central Railway
Bogie Stock
1906

The GCR discontinued building clerestory coaches from 1905. No 957 is a standard 50ft non-corridor six-compartment first. These coaches, built by contractors, made up block trains for London suburban services. The bogies are the standard 8ft wheelbase GCR design. No 957 was withdrawn in 1953 and converted for service use numbered DE 320179.

Nos 664 and 695 are both of the elliptical-roof, dual-brake fitted excursion stock nicknamed 'Barnums'. The design appeared in 1910 and the 'Barnums' were among the first stock built at the GCR's Dukinfield works. The 59ft 11in teak bodies are mounted on steel underframes with 10ft 6in wheelbase bogies. The 'Barnums' owed

GCR First No 957 undergoing restoration. J. S. Parsons.

Restoration going ahead on No 695. The Barnum Coach Trust.

their nickname to the American circus train which had recently toured the country. Their layout was influenced by American practice, with recessed inward-opening vestibule entry doors embellished by long grab rails, saloon interiors and matchboarded lower body panels. No 664, the open third, sat 64 passengers and No 695, a brake third, 32. As built, they were finished in varnished teak. Restoration work is going well on No 695.

No: 957. **Type:** First (1906). **Owner:** Great Central Rly Coach Group. **Location:** Quainton Road.
No: 664. **Type:** Third (1910). **Owner:** Private. **Location:** Bewdley.
No: 695. **Type:** Third Brake (1911). **Owner:** The Barnum Coach Trust. **Location:** Goathland.

Great Southern & Western Railway
Composite Brake — (5ft 3in guage)   1906

In 1906/7 the Great Southern & Western Railway built a number of particularly splendid clerestory corridor coaches to make up the Cork-Rosslare boat train. They were 66ft long, some running on

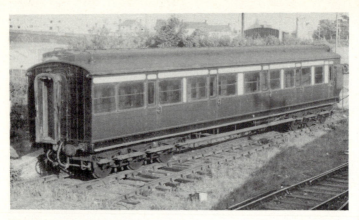

No 861 as restored. Charles P. Friel.

twelve-wheeled bogies. The clerestory itself was wider than customary on English stock. The outside appearance was more impressive than the interiors as the third-class seating was notably cramped. Most of the other GSWR stock was still six-wheeled at that time. Some of the boat train coaches, including the restaurant car, survived into the 1960s.

No 861 was originally designed as a tri-composite with two first, three second-class and two third-class compartments. However, it went into traffic as a first/second composite seating 52. A very spacious brake/luggage compartment is fitted and one point of interest is that the compartments are entered through hinged doors from the corridors. Gas lighting was fitted; fortunately the full fittings remained and have now been restored to use. No 861 is 66ft long, 8ft 6in wide, weighs 36 tons and runs on two six-wheeled bogies. Originally condemned in the early 1960s, it was reprieved for use on the Inchicore (Dublin) Works staff train and numbered 484A. In August 1972 it was again condemned and actually started on its way to the CIE scrap yard, but was rescued at the very last moment. A sales counter has been fitted into part of the van space. No 861 has been restored to its original livery of crimson with off-white panels. It is regularly used on the summer Saturday 'Portrush Flyer' between Belfast and Portrush.

No: 861. **Type:** Composite Brake. **Owner:** Private syndicate. **Location:** Whitehead.

County Donegal/Clogher Valley Railway
Diesel Railcars 1906

The County Donegal survived as long as it did by making extensive use of diesel railcars. The first rail motor vehicle was a four-wheeled petrol-engined inspection car, later railcar No 1, built by Allday & Onions in 1906. It was rebuilt with an enclosed body and new engine in 1920, thereafter looking like a garden shed on wheels. The 1926 coal strike saw it pressed into use for passenger and mails traffic and its success led to the CDR buying further petrol and, later, diesel railcars. No 1, later re-engined, worked until 1956. Railcar No 3 (the second to carry that number) was built in 1926 for the standard-gauge Dublin & Blessington tramway by Drewry Car Co and purchased by the County Donegal in 1932. This is a petrol-engined, double-ended car with tramcar-type bodywork and a four-axle layout. It was demoted to a trailer in 1944 and lasted in service until 1960.

The third preserved railcar was obtained from the Clogher Valley Railway on closure in 1942 and became the second No 10 CDR railcar. It is a 33ft long, 28-seater diesel-engined car with the engine driving a power bogie articulated to the car chassis, and was the first

Petrol railcar No 1, as preserved. Belfast Folk & Transport Museum.

Railcar No 10 (ex CVR), as preserved. Belfast Folk & Transport Museum.

of this type in Ireland. It was built by Walker Bros of Wigan. No 10 was used after the end of public services on demolition work.

CDRJC railcar No 12 was the first power-bogie type of diesel railcar on the County Donegal. It was built at the GNR's Dundalk Works.

No: 1. Type: CDRIC Petrol Railcar (1906). **Owner:** Belfast Transport Museum. **Location:** Belfast.

No: 3. Type: CDRJC Petrol Railcar (1926). **Owner:** Belfast Transport Museum. **Location:** Belfast.

No: 1. Type: CVR diesel Railcar (1932). **Owner:** Belfast Transport Museum. **Location:** Belfast.

No: 12. Type: CDRJC diesel Railcar (1934). **Owner:** North-West of Ireland Rly Soc. **Location:** Londonderry.

Lancashire & Yorkshire Railway
Saloons   1906

This section includes four special-use saloons of modern elliptical roof design. The directors' saloon has the usual layout of centre offices and saloons at each end with large observation windows. It is 45ft long. Withdrawn as BR No 45037, it is now being restored.

No 50 was built as a first-class family saloon at Newton Heath works. It was numbered 10722 in 1923 and became No 995 in 1933. It was withdrawn in 1948 and converted in 1954 for the BR Research Department. The interior has been gutted and a mobile generator set fitted at one end.

The L & Y first possessed a dynamometer car in 1896. The new 50ft car was built at Newton Heath. At one end there is a raised 'balloon' end from which footplate working could be observed. The interior also comprises a recording room, compartment, kitchen and storeroom. Plate-framed bogies were fitted, but more recently replaced. The recording accuracy of the car was suspect and its fallibility confirmed in LMS days; so, in 1929, the car was generally modernised and fitted with a gangway connection at one end. In this condition it was numbered 45050 — Dynamometer Car No 1. It was used during the locomotive exchanges in 1948 and withdrawn in 1971. It was presented to Derby Corporation by BR with full instrumentation.

No 247 was the last L&Y-designed coach to carry a pre-Grouping number. It was used to convey the medical officer around the North Western Division of the LMS. Numbered 10825 by the LMS and later No 45017, it was withdrawn in 1971. It will be restored for private party hire work.

No: (45037). **Type:** Director's Saloon (1906). **Owner:** Midland & GNJ RS. **Location:** Sheringham.

No: 50. **Type:** First Family Saloon (1911). **Owner:** Derby Corporation. **Location:** Butterley.

No: 293. **Type:** Dynamometer Car (1913). **Owner:** Derby Corporation. **Location:** Butterley.

No: 247. **Type:** Medical Officer's Saloon (1923). **Owner:** The Historic Rolling Stock Group. **Location:** Dinting Railway Centre.

L&YR Director's Saloon as preserved. T.J. Edgington.

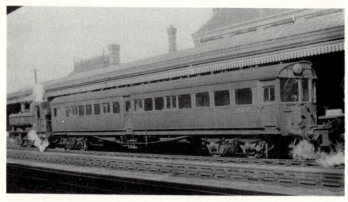

70ft railmotor trailer No 197 in BR days. R.M. Casserley.

To reduce operating costs the concept of a small engine unit and coachwork as one vehicle was taken up by British railways from 1900. Most proved underpowered and unreliable, but the GWR evolved a competent design and came to operate the largest fleet of steam railcars. From 1905 the GWR developed its version of the auto-train, with a conventional locomotive pushing or pulling anything up to four trailers. From 1915-36 the steam railcars were progressively rebuilt as auto-trailers.

No 38 is one of the short 59ft 6in trailers, of which six were built. Seating capacity of these cars was 64. No 92 is a 70-footer — 9ft wide over the mouldings — seating 70 and a gangway connection is fitted at the non-driving end. The windows and hinged ventilator are arranged differently to No 38. Trailer No 212 was originally steam railmotor No 93, built in 1908. This is a 70ft car seating 61 as a railmotor and 77, in three saloons, as a trailer. It belonged to the last series of steam railmotors to be built and was also among the last to be converted as a trailer.

No: 38. Type: Auto-trailer (1907). **Owner:** Private. **Location:** in store.

No: 92. Type: Auto-trailer (1912). **Owner:** Great Western Soc Ltd. **Location:** Taunton.

No: 212. Type: Auto-trailer (1908) (rebuilt 1936). **Owner:** Great Western Soc Ltd. **Location:** Didcot.

Great Western Railway
Toplight Stock 1907

Following the 'Dreadnoughts', the GWR moved to a more conventional design of elliptical-roofed coach which was built in 57ft and 70ft lengths from 1907-22. These were known as 'Toplights' on account of the hammered glass fanlights above the windows. Earlier batches had wooden body panelling with mouldings and gas lighting, while later batches had steel panelling and electric lighting. No 70-footers survive, probably because they were not converted as either service vehicles or camping coaches. The preserved 57ft vehicles have been drawn from these ranks. Generally speaking, the 'Toplights' disappeared by 1958.

No 7538 was obtained as the staff van from Southall breakdown train: it had originally been a brake tri-composite with two compartments for each class. Nos 2426/34 are eight-compartment thirds and were fitted with electric lighting from the start. No 2426 was converted to a camping coach in the late 1950s and is currently in use as a shop at Bewdley. No 3930 was one of the last batch of coaches completed as coaches in the First World War; many of the 1914 orders were turned out as ambulance vehicles. No 3930 was built with steel body panelling and since much of the interior remained when purchased for preservation it will probably be the first to go back into traffic. No 1145 is a steel-panelled van. No 1159 was an ex-ambulance coach rebuilt as a full brake in 1925 and converted to a medical officer's coach in 1944/5. The 57ft Toplights ran on 8ft and 9ft wheelbase bogies, many with the American equalised beam type.

No: 7538. **Type:** Brake composite (1907). **Owner:** Mr W. H. McAlpine/Flying Scotsman Enterprises. **Location:** Didcot.
No: 1145. **Type:** Bogie Brake Van (1922). **Owner:** Private. **Location:** Didcot.

57ft third No 3663 as built.

No: 2434. **Type:** Third (1910). **Owner:** Great Western Soc Ltd. **Location:** Bodmin.

No: 3930. **Type:** Third (1915). **Owner:** Private. **Location:** Bridgnorth.

No: 2426. **Type:** Third (1910). **Owner:** Severn Valley Rly Co Ltd. **Location:** Bridgnorth.

No: 1159. **Type:** Medical Officers Saloon. **Owner:** Great Western Soc Ltd. **Location:** Didcot.

Great Eastern Railway
Corridor Stock 1907

It was not until 1897 that the Great Eastern introduced bogie stock for main line services and the first complete corridor train went into traffic in 1904. Elliptical-roofed corridor coaches were built for the 'North Country Continental' in 1906. A twelve-coach set of similar vehicles was built for the 'Norfolk Coast Express' in 1907; this train ran between Liverpool Street and Cromer with through coaches for Sheringham and Mundesley. From then onwards, more elliptical-roofed corridor stock was built for general service by the GER.

The post-1906 corridor stock was characterised by vertical panels to the bottomsides of the body. Most of these coaches were side-corridor, although open firsts and thirds were also built. The general underframe length was 54ft. The bodies were all-wood.

No 295, from the 'Norfolk Coast' set, became LNER No 62377 and was then converted for service use as No DE 320325. The GER

Similar GER brake third running as BR No E 62429E.

corridor stock had been withdrawn by about 1958/9. No 295 is used for service purposes by the M & GNJRS and has not yet been restored.

No: 295. **Type:** Third Brake. **Owner:** Midland & GNJRS and other groups. **Location:** Sheringham.

Hull & Barnsley Railway
Non-corridor Stock   1907

No 58 as restored 1974. J. Boddy.

The Hull & Barnsley was one of the last major new railways to be built. It opened in 1885, but by 1932 the main passenger service between Hull and Cudworth had been withdrawn and all regular passenger workings finished in 1955. The Hull-Sheffield express workings ran only from 1905-17, so it is of particular interest that two coaches should remain as a memorial to the H&BR.

The earliest (1885) H&BR coaches were four-wheeled, but bogie stock was built for the Hull-Sheffield service from 1907. This was formed in three-coach sets. The bogie stock was 50ft long and 8ft

9½in wide with elliptical roofs. No 40 is a six-compartment brake third, later LNER No 25040 and latterly a signal engineers' van. It was built by RY Pickering. No 58, LNER 25058, is a semi-corridor lavatory brake third and was built by Metropolitan C&W. The original livery of these coaches was varnished teak panels with gold lettering shaded in blue.

No: 40. Type: Third Brake (1907). **Owner:** H&BR Stock Fund. **Location:** Grosmont.

No: 58. Type: Lavatory Third brake (1908/9). **Owner:** H&BR Stock Fund.

Highland Railway
Six-wheel Composite   1908

From 1897 bogie stock was generally introduced for main line services on the Highland Railway. However, six-wheelers such as No 89 appeared as late as 1908. This coach, originally used between Strathpeffer and Inverness, has three third-class and two first-class compartments and a coupé first-class with end windows. This feature was associated with coaches produced under the aegis of David Jones, the locomotive superintendent from 1870-96. No 89, built at the HR's Lochgorm works, Inverness, has the characteristic narrow matchboard body panelling. It was latterly in service stock and nicely restored in 1968 at Lochgorm to the dark green livery

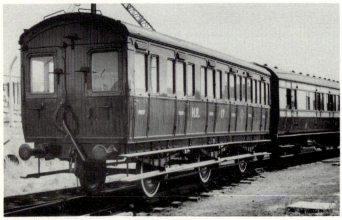

As restored (as No 17) in 1968. W. A. C. Smith.

used after 1903. In the photograph it is numbered 17, but this has now been corrected to No 89.

No: 89. Type: Six-wheel Composite Coupé. **Owner:** Scottish RPS. **Location:** Falkirk.

South Eastern & Chatham Railway
'Birdcage' Stock   1909

After the formation of the SE&CR in 1899 the new organisation did its best to improve the reputation of the old companies by producing some very creditable bogie stock. Among the best-known were the Trio C three-coach non-corridor sets built in the 1900s. These had raised glass lookouts for the guard at each brake end, which has given them the nickname 'Birdcages'. The Kent & East Sussex has two brake thirds from a Trio C set sold to the Army at Longmoor by the SR in 1943. Both are 54ft long and 8ft 10½in wide. No 5312 (the last WD number) is a seven-compartment coach; No 5311 has six compartments with a lavatory. Both coaches came from Longmoor in 1970. The Bluebell has a six-compartment 'Birdcage' brake. It was later SR 3334 and went to departmental stock in 1951. Now largely restored in SE & CR livery, it is in use on Bluebell services. No 3582 is a 50ft family Birdcage saloon brake with the customary layout of a centre vestibule flanked by a saloon on one side and

No 1061 on the Bluebell Railway. D. Hill.

lavatory and servants' compartment on the other. It was converted for service use in 1954, losing its interior in the process.

No: 1061. **Type:** Brake Composite. **Owner:** Bluebell Rly PS. **Location:** Sheffield Park.

Nos: 5311, 5312. **Type:** Brake Third. **Owner:** Tenterden Rly Co Ltd. **Location:** Rolvenden.

No: 3582. **Type:** Brake Composite Family Saloon. **Owner:** South Eastern Steam Centre. **Location:** Ashford (Kent).

London & North Western Railway
TPO Van Brake Third 1909

The TPO van, built at Wolverton, was originally fitted with pickup arms and collection net. It is 57ft long with a wooden body and originally had gas lighting. Until 1940 it ran on the London to Holyhead service, numbered 20 at first, then LNW 9520, then 3227 in 1923 and finally 30244 in 1933. The arms and net were removed in 1945. It was then used on the Bristol-Birmingham, later Bristol-Derby, TPO until 1949. It was withdrawn in 1961 and was later bought by the RPS. At present, external restoration has not been begun and it is used to house the RPS small relics collection.

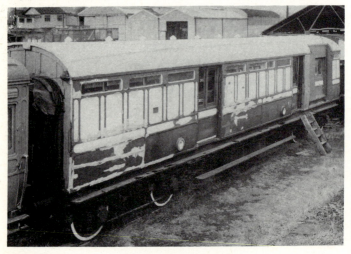

The TPO van as preserved. Railway Preservation Soc.

LNW brake third, as preserved. Railway Preservation Soc.

The brake third has six compartments and was built at Wolverton. It is a typical LNW wooden-bodied non-corridor coach. In service until 1957, latterly as BR No M 22687M, it was sold to the National Coal Board in 1958 for the Hednesford-Rawnsley Colliery miners' service which was withdrawn in 1964. The coach has not yet been restored.

No information is available for the passenger brake van.

No: 20. **Type:** TPO Van (1909). **Owner:** RPS, Chasewater Lt Rly. **Location:** Brownhills.

No: (22687). **Type:** Third Brake (1917). **Owner:** RPS, Chasewater Lt Rly. **Location:** Brownhills.

Type: Passenger Brake Van (1905). **Owner:** L&NW Society. **Location:** Brownhills.

Great Northern Railway
Saloons
1909

These coaches follow the elliptical-roofed outline introduced by Gresley on the GNR in 1905. No 3087, a gangwayed invalid saloon, is 50ft long and 9ft wide. The interior layout as built consisted of a lavatory, compartment, vestibule, main saloon, lavatory, first-class compartment and luggage area. From the start it had electric lighting. Total cost when built was £1,891. It was subsequently converted for departmental use and survived to be repainted in BR blue and grey livery numbered DE 320042. It was withdrawn at Norwich.

Nos 397 and 807, 52ft 6in long and 9ft wide, were built as first-class family saloons costing £1718 each. They were fitted from the start with 8ft 6in wheelbase compound-bolster bogies. The interior layout consisted of a vestibule, lavatory, saloon, large saloon, lavatory, servants' compartment and luggage space. No 397 became LNER No 4397 and was later converted for departmental use, withdrawn as BR No DE 320206. Both No 397 and 3087 may be restored for mainline running in the future.

No: 3087. **Type:** Invalid Saloon (1909). **Owner:** Mr W. H. McAlpine/Flying Scotsman Enterprises. **Location:** Carnforth.

No: 397. **Type:** First Saloon (1912). **Owner:** Mr W. H. McAlpine/Flying Scotsman Enterprises. **Location:** Carnforth.

No: 807. **Type:** First Saloon (1912). **Owner:** Private. **Location:** Bewdley.

No 397 in the state in which it was purchased as BR No DE320206. R. Hudson.

Lancashire & Yorkshire Railway
Non-corridor Stock   1910

No 102 is a five-compartment brake third built at the L&YR's Newton Heath service for local services. One of its distinctive features is that the sides of the brake compartment taper inwards of the end guard's lookouts. The dimensions are 54ft long and 9ft 3½in wide. No other general-service L&YR coaches have as yet been preserved. The coach survived into BR days as M 23964M and was then converted in the early 1950s for service use. Originally it was

No 102 in department stock. R. Higgins.

intended to restore the vehicle to running condition, but it is now likely to remain as a staff mess for volunteer workers on the KWVR.

No: 102. **Type:** Third Brake. **Owner:** L&Y Saddletanks Fund. **Location:** Haworth.

London & North Western Railway
Richmond Line Stock   1910

As already described on page 23, the North London Railway used close-coupled block sets of four wheeled coaches for its passenger services. From 1909 the NLR entered an agreement by which, although remaining independent, it would be worked by the LNWR. This led to electrification of most of the NLR services by 1922. In 1910 the Wolverton works of the LNWR built some trains of modern elliptical-roofed four-wheel stock for the Broad Street to Richmond line. In general appearance they followed standard LNWR non-corridor stock. After electrification of the principal NLR services, there was still work for the four-wheel stock on the lines to Poplar and from Broad Street to Potters Bar. By October 1933 the original NLR stock had been withdrawn, but those of the 1910 Richmond design were made up into three sets of thirteen coaches. No 31, a three-compartment brake third, was sold at some time to the

No 31 as partly restored. R. C. Riley.

Woolwich Arsenal, where the interior partitions were removed. It is now intended that it will run as open saloon and it is in the process of restoration.

No: 31. **Type:** Four-wheel Third Brake. **Owner:** Tenterden Rly Co Ltd. **Location:** Rolvenden.

Metropolitan Railway
Dreadnoughts   1910

In 1906 the Metropolitan and GCR formed the Met & GC Joint Committee to control the non-electrified lines beyond Harrow to Aylesbury and Verney Junction. The GCR had introduced some of the best non-corridor stock of the time on its services using these routes, so the Metropolitan was bound to follow suit. In fact, the 54ft 'Dreadnoughts' (nicknamed like the GWR vehicles after big-gun battleships) were not unlike the GCR coaches, with high elliptical roofs and panelled teak bodies of 9ft 3in width. The first ten 'Dreadnoughts' were rebuilt from 1905-built electric control trailers; No 427 is one of these. By 1923, 92 coaches had been built of the 'Dreadnought' pattern — brake thirds, firsts and thirds formed in five-coach sets. Some of these were later included in the 'T' stock electric multiple-units, but 65 were in service between Baker Street City and Aylesbury in six-coach sets until electrification to Amersham in September 1961. One train was stored and then used for a commemoration run at the Met centenary in 1963; three coaches of this train have been preserved.

No 427 (the Met numbers were retained by LT) is a seven-compartment brake third; No 465, a nine-compartment third; and No 509, a seven-compartment first (first-class withdrawn from 1941). All were originally finished in varnished teak but repainted teak brown after 1945. Now in use on KWVR trains they are numbered 4, 2 and 3 respectively and painted in a two-tone livery.

No: 427. **Type:** Third Brake (1910). **Owner:** Keighley & WVRS. **Location:** Haworth.
No: 465. **Type:** Third (1919). **Owner:** Keighley & WVRS. **Location:** Haworth.
No: 509. **Type:** First (1923). **Owner:** Keighley & WVRS. **Location:** Haworth.

No 427 as restored. Ian G. Holt.

Pullman Car Co
Pullman Cars 1910-23   1910

The Pullmans included in this section are rather a mixed bag. *Emerald* and *Sapphire,* the oldest, were built for the SECR Charing Cross to Dover service by Birmingham RCW and are 57ft 6in cars. *Emerald,* which is in good condition, was used from 1955-9 as a staff training car, latterly as a camping coach. *Sapphire* is currently rather derelict. *Topaz* was built by Birmingham RCW, also for SECR services and to 57ft 6in length. It was used on SR workings until 1960 and then beautifully restored at the Pullman Car Co works at Brighton with the help of a donation from an enthusiast. It is now in the condition in which it ran in the 1920s, with a fully restored interior, and has the original wood-framed bogies. It was subsequently on show at Clapham. *Malaga* is a 63ft 6in body coach which spent all its life on the SR until withdrawn in 1961. It is now Ian Allan's board room and senior staff dining saloon.

Padua and *Rosalind* were built for SE & CR Pullman services by Birmingham RCW. They are all-wooden twelve-wheeled cars, 63ft 6in long and 8ft 7in wide. *Padua* was originally a first-class parlour car, later second-class and finally a guard second. Both survived on the SR until about 1960, when they were converted as camping coaches and were latterly in departmental use. Neither has as yet been restored.

Cars Nos 135/7 were known as Pullman 'J' type, or Clayton Pullmans. Both were built originally as ambulance coaches by the LNWR in 1914. In 1921 Claytons of Lincoln used the underframes for new first-class Pullman kitchen cars: *Elmira* (No 135) and *Maid of Kent* (No 137), both 58ft 6in long and 8ft 6in wide and on eight-wheel bogies. Both became composites in 1933/4 and then, in 1948, third parlour cars. They were then used on the 'Thanet Belle' until 1959 and were converted into camping coaches for the London Midland Region in 1960. Both were acquired in 1968 and are now used as camping coaches for visitors by the R&ER.

Name: *Emerald.* **Type:** First Kitchen (1910). **Owner:** Conwy Valley Railway Museum. **Location:** Bettws-y-coed.

Name: *Topaz.* **Type:** First Parlour (1913). **Owner:** Department of Education & Science. **Location:** National Rly Museum, York.

'Topaz' as withdrawn in 1961.

Name: *Sapphire.* **Type:** First Kitchen (1910). **Owner:** South Eastern Steam Centre. **Location:** Ashford (Kent).

Name: *Padua.* **Type:** Guard Second (1920). **Owner:** Mr W. H. McAlpine/Flying Scotsman Enterprises. **Location:** Carnforth.

Name: *Rosalind.* **Type:** Kitchen First (1921). **Owner:** Mr W. H. McAlpine/ Flying Scotsman Enterprises. **Location:** Carnforth.

Name: *Malaga.* **Type:** First Kitchen (1921). **Owner:** Ian Allan Ltd. **Location:** Shepperton.

Name: Cars Nos 135/137. **Type:** Third Parlour (1921). **Owner:** The Ravenglass & Eskdale Rly. **Location:** Ravenglass.

Great Western Railway
Saloons   1910

No 6479 as purchased from British Railways. David Wharton.

The GWR inspection saloons already described were rebuilt from family and special saloons in the 1930s or later. Most of the District Engineers' saloons were former Bristol & Exeter four-wheelers converted in the 1900s. One exception was No 6479 (Engineer's series), which was built in 1910 using the underframe from Manchester & Milford Rly composite No 149. A new body was provided, 45ft 6in long, giving a saloon at each end and a central 'services' section. The roof was of the three-centre arc pattern. Originally allocated to Taunton, the coach was withdrawn from Gloucester as BR No 80977.

No 9055 was built as a third saloon displacing an older vehicle. A standard 57ft underframe was used and the body generally conformed to contemporary 'Toplight' stock. The coach has the usual saloon-at-each-end layout and a centre side-corridor compartment; it seated 44 as built. Electric lighting was provided from the start. Later in the departmental series as DW 150127 the coach came to the SVR in 1972.

No 9369 is what is known as a nondescript saloon: that is, unclassed and built for party hire. The layout is different from the other type of saloon in having a side corridor giving access to two saloons with inward facing seats, an ordinary compartment, a lavatory and the guard. The body is 57ft long and 8ft 11in wide with steel panelling and no toplights. One of a batch of three, No 9369 had been ordered in 1914. It was originally allocated to Bristol. As with 9055 it ended its days in service stock. (*See also* No 9103, *page 125*). Both Nos 9055 and 9369 are to be restored for private hire work.

No: 6479. **Type:** Inspection Saloon (rebuilt 1910). **Owner:** Dart Valley Rly Association. **Location:** Paignton.

No: 9055. **Type:** Third Saloon (1912). **Owner:** Great Western (SVR) Association. **Location:** Bewdley.

No: 9369. **Type:** Third Saloon (1923). **Owner:** Private. **Location:** Bewdley.

London & North Western Railway
Inspection Saloon   1910

This six-wheeled coach was one of three saloons all now at Bewdley which formed the Longmoor Military Railway officers' train. It was built as an inspection saloon and is believed to have been used by

In service on LMR. G. M. Kichenside.

Mr George Whale, the then Chief Mechanical Engineer of the LNWR. As the photograph shows, it has two main saloons and spacious end verandahs. Its subsequent history is not clear but most probably it came to the Longmoor Military Railway just before or in the early days of the 1939-45 war. In any case, its LMS number was assumed by a new inspection saloon built in 1944 which has also been preserved — *page 179*. In due course it will be repainted in LNWR livery.

No: (45021). **Type:** Inspection Saloon. **Owner:** Transport Trust. **Location:** Bewdley.

London & South Western Railway
Invalid Saloon   1910

This coach of typical LSW outline was built as an invalid saloon. At one end is a luggage compartment, then the main saloon which leads through a short side corridor to a lavatory and separate first- and second-class compartments. The body is 46ft 6in long. Corridor connections were fitted at both ends. It was renumbered 4105 in 1912 and became Southern Railway No 7803, being sold to the Longmoor Military Railway for use as an officers' saloon in 1938. When on the LMR it ran with the SECR family saloon (*page 72*) and LNWR inspection saloon (*page 104*) as the officers' special train. The corridor connections were removed on the LMR and the body ends panelled over. It was sold to the Transport Trust on the closure of the LMR

As SR No 7803. R. C. Riley.

as Army No 3007. The Severn Valley Railway intend to repaint it in LSWR livery.

No: 11. **Type:** Invalid Saloon. **Owner:** Transport Trust. **Location:** Bewdley.

Great Northern Railway
Brake First   1910

The elliptical roof stock so closely associated with Sir Nigel Gresley first made its appearance in 1905 for GNR stock and in the following year for the East Coast Joint Stock. Over 200 general service coaches of this type were built for the GNR between 1905 and 1922, in a variety of small batches. No 3178 is a first brake with five compartments, with body measuring 61ft 6in over the ends and 8ft 6in in width. It was one of a pair, the other, No 3177, was used in the GNR royal train. As the photograph shows, these early Gresley GNR coaches can be distinguished by the toplights to windows and droplights. The coach became LNER No 43178 and was withdrawn from traffic in early BR days to be converted to departmental stock. When found by Mr Kohring it was the packing van (in the break-down train) at Scunthorpe as No DE 320651. It now has 8ft 6in bogies. The intention is to convert it to a club car with one compartment and all exterior doors restored. It will eventually be brought back to varnished teak finish.

No: 3178. **Type:** First Brake. **Owner:** B. G. Kohring. **Location:** Taunton.

As purchased for preservation. A. Turner.

Swansea Harbour Trust
Inspection Saloon 1911

This four-wheel coach was built between 1850 and 1860 by the Manchester, Sheffield & Lincolnshire Railway. At a later date it was sold to a South Wales railway, possibly the Brecon & Merthyr, who sold it at the end of the nineteenth century to the Swansea Harbour Trust. Its new owners rebodied the coach as an inspection saloon with open verandah ends about 1911. It was acquired by the present owner in 1969, having been used as an office and store at Swansea Docks.

Type: Inspection Saloon. **Owner:** Private. **Location:** Caerleon.

The saloon as preserved. G. Crew.

North Eastern Railway
Non-corridor Stock 1911

The NER commenced building elliptical roof stock in 1906 and one of the most familiar types was the 49ft eight-compartment non-corridor third. They were 8ft 6in wide. The last examples were built in 1923. The bodies were of all-wooden construction with bodyside panel mouldings, and the underframes steel with wooden headstocks. Five-a-side seating was provided. The earlier examples — such as 1972 — had gas lighting as built. These coaches were in service until the late 1950s, largely displaced by diesel multiple-units or service closures. The two preserved coaches were sold to the

NER non-corridor stock hauled by Class D20 4-4-0 No 62381 on a York-Goole workmen's train at Selby in 1955. P. J. Lynch.

National Coal Board for workmen's services at Lynemouth and Ellington Collieries in Northumberland. Although intact and sound, considerable work is required to bring them to acceptable condition. It is intended that both gas lighting and Westinghouse air-braking will be restored.

No: 1972. **Type:** Third (1911). **Owner:** North of England Open Air Museum.
 Location: Beamish (Co Durham).
No: 118. **Type:** Third (1913). **Owner:** North of England Open Air Museum.
 Location: Beamish (Co Durham).

South Eastern & Chatham Railway
Non-corridor Stock   1911

As already noted with the ex-LBSC coaches used on the Isle of Wight railways, the stock coming from the mainland was generally rebuilt. From 1948 ex-SECR coaches were sent over, having been converted from 'Trio C' three-coach 'Birdcage' sets. The 'Birdcage' lookouts were removed from the brake coaches and the lavatories taken out; the resulting space was then made into a small saloon compartment in many of the other coaches. Steel panelling was also liberally applied to the post-war transfers to the IOW and all were renumbered into a new series.

No 4145 is a four-compartment brake and No 4149 has three compartments. Both are 54ft by 8ft 10½in vehicles. No 6375, of similar dimensions, was originally a lavatory composite with three first- and four third-class compartments. The central area of the lavatory and two adjacent compartments were converted to form a saloon. Latterly No 6375 ran as second class only although not re-numbered; some similar coaches had been demoted as thirds (seconds) when sent over to the Island. One further note: the Island trains had Westinghouse air braking.

Nos: 4145, 4149. **Type:** Third Brake. **Owner:** Wight Locomotive Soc. **Location:** Haven Street (IOW).
No: 6375. **Type:** Composite. **Owner:** Wight Locomotive Soc. **Location:** Haven Street (IOW).

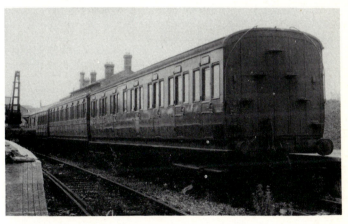

No 6375 as purchased for preservation. Lawrence Dopson.

Great Northern Railway (Ireland) Directors' Saloon (5ft 3in gauge)   1911

Similar to GNR (English) stock of the period, with a wooden-panelled body with square-cornered windows and panel mouldings, it is 48ft long and 9ft 6in wide. The body consists of two saloons with a lavatory and seats 22 passengers. Gangway connections are fitted at both ends. More recently it became UTA and NIR No 150 and had been out of service for some while before being purchased by the UTDA and presented to the RPSI in 1973. Work is now going ahead to restore it to original condition and the interior will be recarpeted in suitable style.

No: 50. **Type:** Director's Saloon. **Owner:** Ulster Tourist Development Association. **Location:** Whitehead.

NSB (Norwegian State Railways)
Four-wheel Coaches   1911

These coaches were bought to accompany the Norwegian State Railways Class 21C 2-6-0 locomotive No 377, *King Haakon 7*, by Mr Gerald Pagano. No 547 was purchased in 1973 and is a verandah-ended clerestory-roofed brake which was last used in Norway as the brake van on timber trains. No 1001, bought in 1971, is a verandah-ended coach with the more modern high-domed roof. Both coaches have teak matchboarded bodies. No 1001 is now completely restored with an oiled and varnished teak finish.

No: BFV 21/547. **Type:** Passenger Brake (1911). **Owner:** Norwegian Loco Preservation Group. **Location:** Loughborough.
No: 1001. **Type:** Coach (1915). **Owner:** Norwegian Loco Preservation Group. **Location:** Loughborough.

Taff Vale Railway
Non-corridor Stock   1912

An unusual vehicle which was built as a seven-compartment third but, following damage in a crash, emerged to be rebuilt by Gloucester RCW as a three-compartment third brake in 1912. In this form it was a 38ft non-bogie eight-wheeler. It ran until 1928 when it was withdrawn and the body sold for use as a farm store. Purchased by the Museum Trust from Bath in 1972.

No: 203. **Type:** Third Brake. **Owner:** Somerset & Dorset Rly Museum Trust. **Location:** in store.

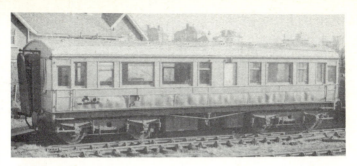

Top: *GNR Director's Saloon No 50 as purchased for preservation.* C.P. Friel.
Centre: *NSB No 1001 in transit.* G. Pagano.
Bottom: *TVR No 203 as built.* Gloucester RCW.

Midland Railway
Royal Saloon 1912

The Midland Railway built a six-coach royal train in 1912, the clerestory coaches being to the design of the carriage superintendent, David Bain. Apart from the royal saloon itself, the other coaches were standard vehicles, the formation being brake first, royal saloon, dining car, saloon No 2934, first, and brake van.

No 1910, the royal saloon, was apparently so numbered as this was the year that King George V came to the throne. It is 59ft 9in over the headstocks and 8ft 9in wide. In general outline it follows the usual Bain features, but is unusual for the wide non-opening windows with louvred ventilators above. Main access is through a wider than normal door midway along the body. The interior consists of an 18ft-long main saloon with a smoking compartment opening off one end and a 'boudoir' off the other. This leads through a side corridor to an end vestibule; at the other end a side corridor leads past the attendant's compartment to another vestibule. The original interior finish was in mahogany and oak wood panelling offset by green upholstery. No 1910 was later used for general VIPs and became special saloon No 2795. In 1933 it was redecorated, still with green furnishings, and given new lighting and forced air ventilation. It was classified a semi-royal saloon, No 809. In this condition it weighed 31 tons, running on standard MR 8ft bogies. In 1951 it emerged from store to be used on the North Wales tourist cruise trains and ran until 1963. Subsequently purchased by the Railway Preservation Society it is now on loan to Derby Corporation and in due course will be restored.

No: 1910. **Type:** Royal Saloon. **Owner:** Railway PS. On loan to Derby Corporation.

No 1910 as built.

London & North Western Railway
Observation Saloon

No 1503 as partially restored. A. MacIntyre.

The number of purpose-built observation coaches in Britain has been fairly small. Three vehicles were built by the LNWR for its scenic lines in North Wales — particularly that from Llandudno Junction to Blaenau Festiniog. These were 50ft coaches of similar general outline to the LNW elliptical-roof corridor stock, but distinguished by glass panels from waist level to just under the roof line at each end of the body. The bodysides had narrow pillars and large windows surmounted by a row of generous hinged ventilators. Two doors were fitted to each side. As the coaches were double-ended, reversible upholstered tram-type seats were provided, seating 72 in all. Each coach was also equipped with a hand brake and vacuum gauge so that it could be worked singly.

The Blaenau Festiniog line was worked by diesel railcars as from 1956, which provided almost equivalent viewing to the LNWR saloons. The latter had come into BR ownership, the preserved example being numbered M 15843M. This was withdrawn about 1963 and bought by the Bluebell. It is a most useful vehicle and has sometimes worked singly — one of the advantages noted above. At present it is partially restored to LNWR livery.

No: 1503. **Type:** Observation Saloon. **Owner:** Bluebell RPS. **Location:** Sheffield Park.

London, Brighton & South Coast Railway
Directors' Saloon 1914

As restored. G. D. King.

The LBSC directors' saloon is one of the finest of its genre, coming from a railway which did not have many large passenger coaches. No 60 was outshopped from the LBSC's Lancing works in summer 1914, built to the design of A. H. Panter, C&W superintendent. The coach is 60ft long and the body itself is only 7ft 8in wide. It runs on six-wheel, 11ft 6in wheelbase bogies. The interior is divided into two saloons either side of central lavatory and pantry facilities. Internally the panelling is in mahogany with attractive moulded ceilings. The larger saloon was originally able to be used as a mobile boardroom with appropriate tables and chairs. Electric lighting is fitted. Observation windows are provided at each end. In original condition the coach was painted in LBSC brown with the railways's coats of arms on each side.

The Brighton elliptical-roofed stock had a limited life as it was out of gauge for the other SR divisions. Fortunately the LBSC directors' saloon had wide route availability, which enabled it to be used over the SR system. At some stage the bodyside windows acquired sliding ventilators. In BR days it was used as an inspection saloon, No DS 291, at first painted red and cream, later green. In 1962 it was used for training drivers for electrified services in Kent and was withdrawn in the mid-1960s. It is now repainted brown.

No: 60. **Type:** Director's Saloon. **Owner:** Bluebell RPS. **Location:** Sheffield Park.

London & North Western Railway
Directors' Saloon 1914

This is a particularly fine directors' saloon, as befits the 'Premier Line.' It is 70ft long and runs on two six-wheel bogies. In general appearance the saloon, original number not to hand, most nearly resembles the fine 2pm 'Corridor' stock of 1908 but does not have recessed doorways. It is gangwayed at each end with observation windows. The interior comprises a centre kitchen/lavatory/guard's section flanked by saloons. The internal finish is particularly attractive, with typical contemporary Adamesque mouldings and panelling which have recently been restored. Electric lighting is fitted. It survived in BR service until the early 1960s. Privately owned, it is now being restored to LNW livery. It is generally used as a buffet car on the M&GN.

No: (45002). **Type:** Director's Saloon. **Owner:** M&GNJR Society (on loan). **Location:** Sheringham.

In BR days as M 45002. T.J. Edgington.

Glasgow & South Western Railway
Corridor Third 1914

This coach is a seven-compartment side-corridor vehicle built by Birmingham RCW. There are four doors to the corridor side. The body (56ft 7in by 8ft 9in) is wooden with full panelling on sides and ends. It was sold out of service by the LMS to the Admiralty for use at Bandeath Naval Depot near Throsk in Stirlingshire and subsequently lost its gangway connections and solebar footboards. Purchased by the SRPS in 1969 and partially restored in 1972, it has regrettably suffered from the attention of vandals, so that extensive work needs to be done to ensure its survival.

No: 731. **Type:** Third. **Owner:** Scottish RPS. **Location:** Falkirk.

Midland Railway
Dining Car   1914

With the construction of the Settle & Carlisle route in 1874, the MR competed with the West and East Coast routes by running Anglo-Scottish trains to Glasgow and Edinburgh. As from 1893 dining cars were introduced to these services and those working to Glasgow were operated as Midland and Glasgow & Southern Western Joint Stock. No 3463 was one of the last clerestory coaches built by the MR, which did not adopt elliptical roof designs until 1917. It is 65ft long and carried on six-wheel bogies. The general outline is of typical Bain pattern. The layout includes an end kitchen/pantry with two dining saloons, which have the characteristic MR arrangement of two windows to each seating bay. The livery was — and is restored as — standard Midland style. Electric lighting was fitted.

No 3463 later became LMS No 165 and survived into BR days. It was lying at Burton-on-Trent in 1954 when an enthusiast persuaded the then BTC to preserve it and restoration, to a very high standard, was completed in 1957, although it is incorrectly lettered as for Midland & GSW Joint Stock. The coach was on display at Clapham.

No: 3463. **Type:** Third-class Dining Car. **Owner:** Department of Education & Science. **Location:** National Rly Museum, York.

Great Southern & Western Railway (Ireland)
Corridor Coach (5ft 3in gauge)   1914

A number of wooden-bodied coaches such as No 4012 lasted long enough on the CIE to acquire the distinctive black and brown livery introduced in the early 1960s. It is a 57ft long side-corridor second with seven compartments. The body width is 9ft 2in. It is typical in

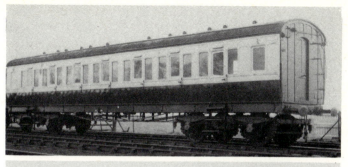

Top: *G&SWR Corridor Third as partially restored.* J. R. Hume.
Centre: *Midland Dining Car as restored.*
Bottom: *GS&WR (Ireland) No 4012 as purchased for preservation (with boards over broken windows).* C. P. Friel.

general design of practice in both Britain and Ireland in the 1910-4 period, with long lights on the corridor side and doors to each compartment on the other side. It was purchased by a group of members in late 1973 and despite broken windows and damaged upholstery is at least intact and in fair overall condition. In due course it will be fully restored.

No: 4012. **Type:** Second. **Owner:** Private (RPSI members). **Location:** White-head.

Caledonian Railway
Non-corridor Stock   1915

This coach is a standard 57ft wooden-bodied non-corridor third, 9ft 4in in width, seating 108 passengers. The original number is uncertain, it is believed to have been either No 426 or 969. It was sold by the LMS for use at the Admiralty depot at Bandeath and purchased from there by the SRPS in 1970. It has not yet been restored.

No: See above. **Type:** Third. **Owner:** Scottish RPS.. **Location:** Falkirk.

The coach at Bandeath. W. S. Sellar.

LNWR
Electric Motor Coach   1915

The first LNWR electrified service was that from Willesden Junction (High Level) to Earl's Court, inaugurated in 1914 at 630v dc with third-rail collection. By 1922 the London area electrification of the

Oerlikon multiple units on the Broad Street-Richmond service.

LNWR was complete apart from the Rickmansworth branch, converted in 1927. For these services, the LNWR provided about 80 three-coach multiple-unit electric sets, all of open layout with hand-worked sliding doors and gangways between each coach of a three-car train. The first sets were built in 1914, later to be converted for ac overhead traction between Lancaster and Heysham. Two batches of what became known as the Oerlikon stock were built — in 1915 and between 1921-3. The stock was built by Metropolitan C & W and Wolverton Works, the electrical equipment being supplied by Oerlikon. The cars all had large bodyside windows with hinged ventilators above and end vestibules. The motor cars, seating 48, were 57ft in length and 8ft 11½in wide, the internal layout including longitudinal and transverse seats.

The Oerlikon stock worked on all the London area LNW section services until withdrawals began in the 1940s. In 1957/8 new compartment stock was brought into service to replace them but the last Oerlikon sets, including motor car No 28249, remained in traffic until early 1960. No 28249 was subsequently used for some time at Stonebridge Park car depots for shunting and has not been seen on public display since being preserved.

No: (28249). **Type:** Motor Third Brake. **Owner:** Department of Education & Science. **Location:** In store.

North British Railway
Invalid Saloon
1919

As restored, 1974. A. Harper.

This wooden-bodied invalid saloon was built at Cowlairs Works in 1919. The centre of the coach was provided with a kitchen and guard's compartment. It ran as a composite in LNER days, numbered 32283, and in 1951 was rebuilt with observation windows at each end and a large kitchen as a district engineer's saloon No 320577. Purchased by the SRPS in 1972, it has now been restored to NBR livery and has one end panelled over and fitted with a corridor connection. It is used on railtours as an observation car, buffet and sales coach.
The vehicle is 49ft 8⅛in long and 9ft wide.

No: 461. **Type:** Invalid Saloon. **Owner:** Scottish RPS. **Location:** Falkirk.

Great Eastern Railway
Inspection Saloons
1920

No 1 was built at Stratford Works as an inspection saloon for the GER's American general manager, Sir Henry Thornton. In the tradition of American whistle-stop tours it was equipped with a handsome verandah platform at one end with ornamental ironwork.

The interior consists of a saloon at the verandah end, central services area with kitchen and a large dining saloon. The exterior follows GER practice at the time with elaborate panelling and vertical mouldings on the body bottomsides. No 1 is 55ft long over the body ends with a corridor gangway at one end. The body is 8ft 9in wide. In 1934 Gresley light-type compound bolster bogies were fitted. In LNER days the coach was numbered 61 and it continued to be used in the Great Eastern area for railway officials and VIPs. In 1948 it moved to York where it was used as the saloon for the North Eastern Region general manager and renumbered DE 962450. It was acquired in 1971 and beautifully restored at BREL's York works to GE varnished teak livery and to main line running condition.

No 63 is an elliptical-roofed inspection saloon of similar general outline. It was latterly BR No DE 900271 and withdrawn at Newcastle in BR blue/grey livery. Some of the interior remains. It will probably be retained for internal use by the owner.

No: 1. Type: General Manager's Saloon (1920). **Owner:** Mr W. H. McAlpine/ Flying Scotsman Enterprises. **Location:** Carnforth.

No: 63. Type: Inspection Saloon (1912). **Owner:** Mr W. H. McAlpine/Flying Scotsman Enterprises. **Location:** Carnforth.

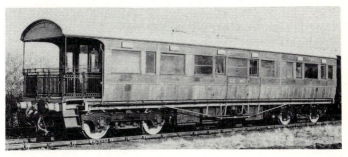

No 1 as restored. T. Boustead.

Midland Railway
Brake Third   1921

The last carriage superintendent on the Midland Railway was R. W. Reid, who introduced elliptical-roofed stock to the railway in 1917. These were plain, unexceptional vehicles the general features of which were followed in the early LMS stock. The third brake is a non-

Top: *MR Brake Third as purchased for preservation.* J. B. Radford.
Centre: *GWRS No 3755 in service in 1958.* R. M. Casserley.
Bottom: *CR Nos 3339 and 7369 reunited in 1974.* A. Harper.

corridor coach with six compartments. Its last days in BR service were as a mobile test laboratory. It is now being overhauled and restored to original condition.

No: (DM 395525). **Type:** Third Brake. **Owner:** Derby Corporation. **Location:** Butterley.

Great Western Railway
Main Line & City Stock   1921

The GWR decided to build bogie stock for its suburban services through to the City in 1913, but the onset of war deferred their construction until 1920/1. Six close-coupled six-coach sets were built and the coaches were only 48ft in length and 8ft 5½in wide with steel-panelled bodies. The general appearance was rather austere and antique, but they were otherwise similar to other 'Toplight' non-corridor stock. The sets were used on City services until 1939 and on other Western Region London suburban workings until 1956/7. A few, including No 3755/6, were transferred to a South Wales miners' train working in 1958 and were in use, with gas lighting in place of electric, until late 1964. Nos 3755/6 have six compartments and a short brake/luggage area.

Nos: 3755, 3756. **Type:** Third Brake. **Owner:** Private. **Location:** Didcot.

Caledonian Railway
Corridor Stock   1921

The Caledonian Railway produced some fine 64ft twelve-wheeled elliptical-roofed stock for its Glasgow to Aberdeen service in 1905 — the 'Grampian' coaches — which were among the most modern of their type. Nos 3339/7369 (LMS numbers) are of the 57ft version (9ft 4in wide) of the 'Grampians.' The third has eight compartments and the brake composite two first-class and four third-class. Both were built at St Rollox works. In early 1958 they were restored to CR livery by BR Scottish Region for special working with Caledonian 4-2-2 No 123. They were used for a number of railtours and special workings until the mid-1960s, when No 3339 was sold to the Bluebell RPS and No 7369 to the SRPS. No 3339 worked on Bluebell trains until September 1974, when it returned north of the Border to join the other coach. No 7369 has been used for railtours by the SRPS and it is the intention that No 3339 will be similarly employed in 1975.

No: LMS 3339. **Type:** Third. **Owner:** Scottish RPS. **Location:** Falkirk.
No: LMS 7369. **Type:** Brake Composite. **Owner:** Scottish RPS. **Location:** Falkirk.

LMS (Northern Counties Committee) Corridor Stock (5ft 3in gauge)   1922

In 1903 the Midland Railway of England acquired the Belfast and Northern Counties Railway and subsequently a managing committee, the Northern Counties Committee, was appointed to run the system. The Midland influence and later, after Grouping, that of the LMS was very strong in the design of the NCC locomotives and rolling stock. By 1922, standard MR/LMS type corridor stock was being built for the NCC and similar vehicles were built into the 1930s. No 243 is a 57ft eight-compartment third of a type almost identical to LMS side-corridor vehicles of the 1924-9 period. Some of these coaches were destroyed during the 1941 'blitz' on Belfast and the LMS transferred stock from Britain to replace them. A number of the NCC coaches were rebuilt for working with diesel multiple-units and many received steel panelling to modernise their appearance. No 243 remained in locomotive-hauled stock until 1975. Nos 238/41 are 1922-built MR 57ft eight-compartment thirds transferred in 1941 by the LMS. They too were in service until 1975. All three coaches had been renumbered in 1959.

Early in 1975 the RPSI purchased ten coaches as a bulk purchase from Northern Ireland railways — NCC and GNR vehicles — from

NCC 1922 Corridor Stock. W. G. Sumner.

more than twenty being withdrawn. They will be used on RPSI railtours and the 'Portrush Flyer.'

Nos: 238 (340), 241 (342). **Type:** Third (1921). **Owner:** Railway Preservation Soc of Ireland. **Location:** Whitehead.
No: 243 (358). **Type:** Third (1924). **Owner:** Railway Preservation Soc of Ireland. **Location:** Whitehead.

Great Western Railway
Bow-Ended Stock   1923

The standard passenger stock introduced by the GWR after Grouping was a steel-panelled vehicle generally of similar lines to the 'Toplight' type (*page 91*). The design had a bow-ended body 58ft 4½in over the ends and usually 9ft wide. Doors to each compartment were provided on both corridor and compartment side. Except for the earliest vehicles, this stock used the newly developed 7ft wheelbase bogie. The 58ft 4½in bow-ended stock was built from 1925-9 (the very earliest coaches were 57ft and flat-ended) and a similar design was built to a 61ft 4½in length from 1929-33. The last of both types was withdrawn in 1963, but a large number were then converted for service use.

No 7976 is of the early flat-ended 57ft type and retains other Toplight features — 9ft bogies and scissor-type gangways. It has two first and four third-class compartments and was purchased in surprisingly complete condition. The other coaches are of standard bow-ended type. No 9103 is a nondescript brake saloon of similar layout to No 9369 (*page 104*). No 1184 is one of the later 61ft 4½in design of the type with 'bulging' 9ft bodysides. No 5539 is one of the non-corridor bow-ended coaches. The SVR also purchased a third, No 4886, for spares only.

Corridor
No: 7976. **Type:** Brake Composite (1923). **Owner:** Private. **Location:** Didcot.
No: 4553. **Type:** Third (1925). **Owner:** Great Western Soc Ltd. **Location:** Didcot.
No: 6045. **Type:** Composite (1928). **Owner:** Private. **Location:** Bewdley.
No: 5085. **Type:** Third (1928). **Owner:** Great Western Soc Ltd. **Location:** Didcot.
No: 5136. **Type:** Third Brake (1929). **Owner:** Severn Valley Rly Co Ltd. **Location:** Bridgnorth.
No: 9103. **Type:** Third Saloon (1929). **Owner:** Private. **Location:** Bridgnorth.
No: 1184. **Type:** Passenger Brake Van (1930). **Owner:** Great Western Soc Ltd. **Location:** Didcot.

No: 814. **Type:** PO Sorting Van (1940). **Owner:** Great Western Soc Ltd.
  Location: Didcot.

Non-corridor
No: 5539. **Type:** Third Brake (1928). **Owner:** Dean Forest RPS. **Location:**
  Parkend.

South Eastern & Chatham/Southern Railway
Non-corridor Stock   1923

The earlier non-corridor SECR stock has already been described
(*page 108*). The three thirds on the Bluebell are not strictly speaking
SECR coaches: the first two entered service at the time of the Grouping,
while No 1050 is a peculiarity, having been made up from odd
compartments left over from conversion of steam stock for electric
sets mounted on a new underframe. All three coaches have ten
compartments seating 100 in dubious comfort. The interiors are
undistinguished, with plain tongue and groove boarding. Length
over the bodies is 60ft with a width of only 8ft 9in: the limited width
meant that they were clear to run over all the SR system. The body
panelling is steel. All three coaches were fitted for working in push-
pull trains: in 1960, 971 was at Horsham, 1050 at Seaton and 1098 at
Lymington. All were withdrawn by late 1962 and arrived on the
Bluebell in early 1963.

Nos: 971, 1098. **Type:** Third (1923). **Owner:** Bluebell RPS. **Location:** Sheffield
  Park.
No: 1050. **Type:** Third (rebuilt 1927). **Owner:** Bluebell RPS. **Location:** Shef-
  field Park.

Southern Railway
Corridor Stock   1923

The SR inherited modern corridor coach designs from the LSWR
and SECR, the best features being combined for the SR's standard
types. The LSWR type known as 'Ironclad' was built on a 57ft
underframe and was a modern-looking steel-panelled design introduced in 1921. 'Ironclads' were built up until 1926 by the SR and
were withdrawn from 1956. No 3204, a four-compartment brake
third, is one of the later SR-built vehicles with standard 8ft bogies. It
was converted for service use in the Exeter district around 1960.

No: 3204. **Type:** Third Brake. **Owner:** Somerset & Dorset Rly Circle. **Location:**
  Taunton.

126

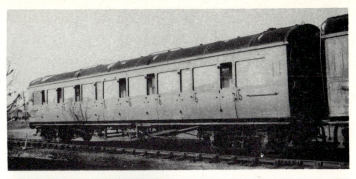

Top: *GWR No 5085 as purchased for preservation*. J. Hosegood.
Centre: *One of the SECR Thirds on the Bluebell*. J.A.M. Vaughan.
Bottom: *A similar 'Ironclad' brake third in departmental stock*. J. Scrace.

Pullman Car Co
Pullman Cars

A Pullman service was introduced on the Great Northern section of the LNER in 1923 between London, Harrogate and Newcastle. For this working two trains of Pullman cars were required: some were the first of a new design 63ft 10in long, 8ft 7in wide, running on a new 10ft wheelbase design of bogie. Up to 1928 new Pullman services were introduced on the LNER from London to Sheffield (later withdrawn) and Bradford, while the Newcastle train was extended to Edinburgh. Later in 1928 the Edinburgh train was extended to Glasgow and became the 'Queen of Scots' Pullman. After 1946 the Pullman trains returned and from 1960 received new stock displacing older cars for use on the Southern Region.

Car No 54 and *Fingall* are examples of earlier cars, the former built by Claytons, the latter by Birmingham RCW. Car No 54 was converted to a brake third in 1937 and both were withdrawn in 1963.

Ibis was one of a batch built by Birmingham RCW for use in the UK, but was purchased before going into service by the Wagons-Lits Company for their Milan to Cannes services. The coaches returned to Britain in 1928.

Minerva and cars Nos 35/36/64 were used on SR Continental services, being built by Midland C&W and Birmingham C&W. Nos 35/36 became second-class cars in 1946. No 36 was later used on the 'Devon Belle.' *Minerva* was originally a kitchen first. These three were remodelled in 1951 to match the post-war Pullmans on the 'Golden Arrow' (*page 196*). No 64 was originally a second-class restaurant car.

The other cars listed were all-steel cars — all except cars Nos 83/4 — built in 1928 by Metropolitan-Cammell, nearly all for the 'Queen of Scots' Pullman. Until then, all-steel main-line coaches were rare in Britain. Cars Nos 83/4 were also used on the LNER services, but were built by Birmingham RCW. *Zena* was originally used on the short-lived 'Torquay Pullman' on the GWR and then went to the Southampton Ocean Liner services. The cars working out of Kings Cross were largely displaced by new Pullmans in 1960-1 and came to the Southern Region for the 'Bournemouth Belle' working until 1967.

The names in brackets are not the original Pullman ones, but those bestowed by the owners who have preserved them. The Bulmer cars are painted dark green and cream in Pullman style.

Name: Car No 54. **Type:** Third Kitchen (1923). **Owner:** Birmingham Rly Museum. **Location:** Tyseley.

Name: *Fingall.* **Type:** First Kitchen (1924). **Owner:** Wight Locomotive Soc (on loan). **Location:** Haven St (IOW).

Name: *Ibis.* **Type:** First Kitchen (1925). **Owner:** Birmingham Rly Museum. **Location:** Tyseley.

Name: *Minerva.* **Type:** Guard First Parlour (1927). **Owner:** Lytham Motive Power Museum. **Location:** Lytham.

Name: Car No 79. **Type:** Third Brake Parlour (1928). **Owner:** North York Moors Rly. **Location:** Grosmont.

Name: *Phyllis.* **Type:** First Kitchen (1928). **Owner:** South Eastern Steam Centre. **Location:** Ashford (Kent).

Name: *Lucille.* **Type:** First Parlour (1928). **Owner:** South Eastern Steam Centre. **Location:** Ashford (Kent).

Name: *Ione.* **Type:** First Kitchen (1928). **Owner:** Birmingham Rly Museum. **Location:** Tyseley.

Name: Car No 75. **Type:** Third Parlour (1928). **Owner:** Ind Coope Ltd. **Location:** 'Spot Gate' PH, Hilderstone.

Name: *Agatha.* **Type:** First Parlour (1928). **Owner:** Wight Locomotive Soc (on loan). **Location:** Haven St (IOW).

Name: Car No 35. **Type:** Third Parlour (1926). **Owner:** Wight Locomotive Soc (on loan). **Location:** Haven St (IOW).

Name: Car No 36 *(Morella).* **Type:** Third Parlour (1926). **Owner:** H.P. Bulmer Ltd. **Location:** Hereford.

Name: Car No 64 *(Christine).* **Type:** Second Restaurant (1928). **Owner:** H.P. Bulmer Ltd. **Location:** Hereford.

Name: Car No 76 *(Eve).* **Type:** Third Parlour (1928). **Owner:** H.P. Bulmer Ltd. **Location:** Hereford.

Name: Car No 83 *(Prinia).* **Type:** Third Parlour (1931). **Owner:** H.P. Bulmer Ltd. **Location:** Hereford.

Name: Car No 84 *(Lorna).* **Type:** Third Parlour (1931). **Owner:** Private. **Location:** Haworth.

Name: *Zena.* **Type:** First Parlour (1928). **Owner:** Keighley & Worth Valley RPS. **Location:** Haworth.

Car No 76 (Eve) as restored. K. P. Lawrence.

District Railway
Electric Stock

The original District Railway electric stock in 1905 was of wooden construction, the first steel multiple-units being delivered in 1920. To replace the earliest moter cars, 50 largely steel cars were built by Gloucester RCW in 1923 known as the 'G' class. These were slab-sided clerestory-roofed vehicles with two pairs of hand-operated doors to each bodyside. Each car was powered by two 240hp motors. The 'G' class motor cars worked in sets with earlier wooden trailer cars. New traction motors were fitted in 1928 and they were also equipped with new control gear. As part of the large pre-war LT re-equipment the 'G' class cars were fitted with air-operated doors and electro-pneumatic brakes in 1938/9. They were then formed in sets with new 'Q' class trailers. From 1962 a number were withdrawn and some others converted to trailers. However, some 'G' stock cars survived until the arrival of the new 'C69' stock and the last ran in traffic in September 1971. They were the final link with the American outline characteristic of the District Line from the earliest electric days.

The 'N' class trailer, No 08063, was built by Metro-Cammell in 1935 for the increased Metropolitan Line service from Hammersmith to Barking. One of a batch of 26, it was equipped from the start with air-operated doors although later it had hand-operated doors for a

No 4184 as preserved. Gloucester RCW.

while when on the District Line. This car also survived in use until 1971.

No: 4184. **Type:** Third Motor Car 'G' Stock. **Owner:** Gloucester RCW CO. **Location:** Gloucester.

No: 4248. **Type:** Third Motor Car 'G' Stock. **Owner:** London Transport Museum. **Location:** Syon Park.

No: 08063. **Type:** Third Trailer 'N' Stock (1935). **Owner:** London Underground Rly Soc. **Location:** Ashford (Kent).

Great Western Railway
Vale of Rheidol Stock (1ft 11⅝in gauge)   1923

Open (LH) and closed (RH) stock in BR days. B. A. Butt.

The original passenger stock supplied for the opening of the Vale of Rheidol Railway was all withdrawn by the GWR by 1938. This stock comprised both open and closed coaches, which were 32ft bogie vehicles. The first stock supplied by the GWR was in the shape of four 32ft coaches, open above the waist with end doors and reversible seats for 48. These are Nos 4997-5000 of 1923. Three similar coaches built on V of R frames, Nos 4149-51, were produced at Swindon in 1938. All these coaches have wire mesh screens up to shoulder height and canvas screens which can be let down from the roof in wet weather. The original closed coaches were replaced by nine coaches in 1938: two brake and seven ordinary. These had new underframes, but reconditioned bogies. The 6ft-wide bodies are of neat steel-panelled design on wooden frames. There are only three doors to each side with low-height wooden seats inside. The brake coaches seat 48, the others 56. No heating or lighting is fitted, as the line works only in summer. The coaches have always been painted in GWR or BR main-line stock livery and are now in standard 'Rail Blue.' For a time, in the 1960s, however, a dark green livery was used. The coaches are now numbered M 4997W, etc, as well as in a local series.

There is also one small four-wheeled passenger van, No 137 built in 1938. No 136, its companion, has been withdrawn.

Nos: 4997-5000, 4149-51. **Type:** Open Stock. **Owner:** British Railways Board. **Location:** Aberystwyth.
Nos: 4995/6. **Type:** Closed Brake. **Owner:** British Railways Board. **Location:** Aberystwyth.
Nos: 4143-8, 4994. **Type:** Closed. **Owner:** British Railways Board. **Location:** Aberystwyth.

Southern Railway
Continental Stock   1924

New Pullman cars were introduced on SECR Charing Cross to Dover boat trains in 1920. To work with them, the SECR chief mechanical engineer, R. E. L. Maunsell, designed some unusual and modern

In 1965 before complete restoration. R. C. Riley.

corridor stock using a 62ft underframe. The general layout introduced some new features, in particular that access was by end doors only without doors to each compartment. Another innovation was that buckeye couplers were fitted to the coaches, together with Pullman gangways. The outer ends of the brake coaches had no gangway connections and screw couplings were provided. Seating in the first class was two-a-side, three-a-side in the seconds. The coach bodies had the flattened elliptical roof used by the SR and matchboard panelling up to waist level. The 'Continental' stock was built by Birmingham RCW, appearing in different lots in 1921/4/7. All but two coaches were withdrawn at the start of the Kent Coast electrification in 1959. No 3554, a seven-compartment brake third, survived until 1961 and was eventually purchased in 1963. It has now been thoroughly and excellently restored to early SR sage green livery.

No: 3554. **Type:** Third Brake. **Owner:** Vintage Carriages Trust. **Location:** Haworth.

London & North Eastern/North Eastern Railway
Open Third   1924

As purchased for preservation. P. Mulholland.

A standard elliptical-roof design for corridor stock was introduced for ECJS stock in 1906 to a design of H. N. Gresley. The NER built its first vestibuled train-sets for its own services from Newcastle to Liverpool and Leeds to Glasgow in 1908. These were on 52ft underframes, with bow-ended bodies in which the roof sloped down at each body end. Windows were fitted on either side of the gangway at each end. Large bodyside windows were a feature of this design and in comparison with the GNR Doncaster design the windows and body mouldings had rounded corners. This type continued to be built up to the Grouping and by the LNER until 1925, although later coaches did not have the end windows.

No 945 (later 2945) is an open third seating 42 passengers in two saloons and was one of two built after the Grouping. It runs on Gresley double-bolster bogies. Withdrawn from passenger service in 1958, the coach was transferred to departmental use as DE 320716. As purchased it was in good condition, with full brake gear; some seats and the interior panelling are intact. It is at present used as a volunteer mess coach, but will eventually be restored for passenger use.

No: 945. **Type:** Open Third. **Owner:** North Eastern Rly Coach Group. **Location:** Goathland.

London & North Eastern Railway
Standard Corridor Stock 1924

As already noted, the standard elliptical roof corridor coach for the ECJS appeared in 1906. This design, with bowed ends, teak body and buckeye couplers was built for ECJS services and for GNR stock from 1905-23. The standard LNER corridor coach design was finalised in late 1923, using a 60ft underframe — and some for use on the Great Eastern were on 51ft underframes. The LNER standard coach was in advance of the other Big Four by virtue of the Pullman gangways and buckeye couplers. The wooden teak-panelled body with squared mouldings and windows was more traditional than

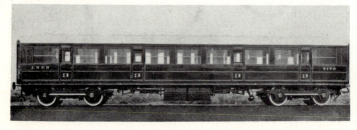

Standard 61ft third No 4173 as built in 1931.

modern, particularly as the LNER persisted with this construction until 1942. In fact there were few differences in design over the 1923-42 period, apart from an increase in body width to 9ft 3in after 1927. All these standard coaches were mounted on the excellent Gresley double-bolster 8ft 6in bogies. The general service stock was withdrawn by 1965.

Nos 1008 (BR 12041) and 4474 (BR 12048) are early eight-compartment thirds, the latter not strictly 'preserved' for restoration. No 1065, latterly 18033, is a composite with two and one coupe first compartments and five third class. It was purchased in 1974 as a mess and tool van. No 24068 (BR 10078) has two first-and four third-class compartments and is intended for early return to passenger service. No 57451 (BR no 16520) is a six-compartment brake third purchased straight from service and therefore complete. No 41384 (BR 16076), a five-compartment brake third, requires considerable work before it can be used. None of the design introduced from 1930 with entry through end vestibules only has yet been preserved.

No 16547 was built as LNER No 43567 and is an open brake third with one saloon seating 32. It was donated in 1974 to the Group by the Tees and Hartlepool Port Authority, who had been using it for training their locomotive drivers. It will be restored to BR maroon livery.

The Brake Third preserved by the 7597 Group is a GE section 52ft 6in body vehicle.

No: 1065. **Type:** Composite (1924). **Owner:** 7597 Group. **Location:** Chappel.
No: 1008. **Type:** Third (1924). **Owner:** Scottish RPS. **Location:** Falkirk.
No: 4474. **Type:** Third (1925). **Owner:** David & Charles Holdings Ltd. **Location:** Newton Abbot.
Type: Third Brake (1927). **Owner:** 7597 Group. **Location:** In store.
No: 43567. **Type:** Open Third Brake (1935). **Owner:** Gresley BSO Group. **Location:** Grosmont.
No: 24068. **Type:** Brake Composite (1937). **Owner:** Private. **Location:** Bewdley.
No: 41384. **Type:** Third Brake (1938). **Owner:** Quainton Rly Soc Ltd. **Location:** Quainton Road.
No: 57451. **Type:** Third Brake (1940). **Owner:** The Gresley Soc. **Location:** Haworth.

London & North Eastern Railway
Quadart Stock   1924

Sir Nigel Gresley, LNER Chief Mechanical Engineer from 1923-41, is well known for his use of articulation in passenger stock. Gresley

Set No 74 in BR days from the non-brake end.

first introduced articulation in 1907 using converted vehicles, but in 1911 he produced some articulated suburban stock for the GNR. These were articulated pairs and were later rebuilt as four-coach ('Quadart') units. The emphasis on this stock for the Great Northern London suburban services was for high seating capacity. From 1921-9 further 'Quadart' compartment stock was built for the GN services with an eventual total of 97 GN and LNER sets. Such sets were made up of teak-bodied coaches with the later sets riding on 8ft wheelbase Gresley bogies. Some other 'Quadart' suburban sets were built for the Liverpool Street-Hertford service.

Two 'Quadarts' formed an eight-coach train. Because of their high seating capacity they were popular with the operators and some continued in traffic until April 1966. The set at Sheringham is one of the early LNER 'Quadarts' built to GNR design, noticeable by the lack of ventilators to the doors and typical GN-style guard's lookouts. It comprises a second brake (six compartments), second (seven compartments) and two first/second composites (seven compartments). Second-class was discontinued after 1938 and the sets became all-third from 1941. From the brake end, the coaches of set No 74 were numbered 48941-4; after 1946, Nos 86272-5. The set has been repainted in a teak brown livery and at present provides the mainstay of the M&GNJR Society train service.

No: 74. **Type:** Quadart Set. **Owner:** M&GNJR Soc. **Location:** Sheringham.

London Midland & Scottish Railway
Standard Corridor/Non-Corridor Stock   1924

The first LMS carriage superintendent was R. Reid from the MR and the general features of Midland practice were adopted for the early LMS standard corridor stock from 1923-9. These coaches were 57ft long and 9ft 3in wide (early coaches only 9ft 1½in) and of all-wooden construction. The coaches had scissors gangways and screw couplings — the LMS, like the GWR, never adopted Pullman gangways and buckeye couplers. A 9ft wheelbase bogie was used. All this stock was withdrawn by 1964.

The third on the Bluebell, used for volunteer sleeping accommodation, has eight compartments with outside doors to each, a type which closely followed post-1917 MR designs. Open thirds Nos 8200/7 are of the Midland 'two-windows'-to-each-bay type seating 56 in two saloons. No 8207 was reserved for preservation by the BRB and is on loan to the Princess Elizabeth Locomotive Society. No 7991 (originally 5682) is of similar layout but of all-steel construction, built by Metropolitan C&W and came to the Severn Valley from workmen's use on the Manchester Ship Canal railway. No 3322 (WD number) was an

An open third of the same design as Nos 8200/7.

open third similar to Nos 8200/7, but converted for ambulance use in World War II and actually captured by the Nazis. It was never de-requisitioned and will now be converted to a first-class saloon. DM 395222 was originally a 57ft first-class coach with three side-corridor compartments and a dining saloon seating 18. Each side has seven large windows with glass ventilators above. It is from the batch Nos 1028-36 (1933 numbers). It was withdrawn in 1955 and converted to an inspection saloon the following year. Acquired by Derby Corporation in 1971 it is now being repaired and overhauled. No 20165 is the only preserved example — so far — of a wooden-bodied LMS non-corridor coach. It has six compartments and is 57ft long and 9ft 3in wide. It is used as the mess room and sleeping quarters for workers on the narrow-gauge R & ER.

Side Corridor
No: (DM 395584). **Type:** Third. **Owner:** Bluebell RPS. **Location:** Sheffield Park.

Open
No: 7991. **Type:** Third (1926). **Owner:** The Historic Rolling Stock Group. **Location:** Bewdley.
No: 8200. **Type:** Third (1927). **Owner:** Southport Loco & Transport Museum Soc. **Location:** Southport.
No: 8207. **Type:** Third (1927). **Owner:** British Railways Board. **Location:** Ashchurch (on loan).
No: (3322). **Type:** (Third) (1929). **Owner:** Cranmore Rly Co Ltd. **Location:** Cranmore.
No: (DM 395222). **Type:** Inspection Saloon (1930). **Owner:** Derby Corporation. **Location:** Butterley.
No: 20165. **Type:** Non-corridor Third Brake (1927). **Owner:** Ravenglass & Eskdale RPS Ltd. **Location:** Ravenglass.

Southern Railway
4-SUB Motor Coach   1925

For the major part of the rolling stock used for the Southern third-rail suburban electrification the SR adapted pre-Grouping steam coaches and mounted them on new underframes. However, 55 three-coach sets were built new in 1925/6 by contractors. These had wedge-ended motor coaches and all vehicles had straight-sided steel-panelled wood-framed bodies. Sets 1285-1310 (later 4300-25) were for Western Division services, Nos 1496-1524 (later 4326-54) were allocated to the Eastern Division. From 1945-6 each set was made up to four cars by the inclusion of a new ten-compartment trailer;

1925/6 4-SUB set No 4332. J. N. Faulkner.

total seating was then 350. The 1925/6 trailers were composites. All 55 sets were withdrawn by early 1962, but the motor coaches from set 4308 were subsequently a de-icing unit.

The motor coaches originally had seven compartments. The interiors of these vehicles were notable for their spartan decor of grained wooden board panelling. However, Nos 8143/4 are an important reminder of the Southern suburban electrification.

No: 8143/4. **Type:** Motor Third Brake (ex Set 4308). **Owner:** Department of Education & Science. **Location:** In store.

Ashover Light Railway
Saloon Coaches — (1ft 11⅝in gauge) 1925

The Ashover Light Railway (in North East Derbyshire) was the last major narrow gauge railway to be built and opened in 1925 largely for stone quarry traffic. A passenger service was provided until 1936. At first four bogie tramcar-type vacuum-braked coaches, built by Gloucester C&W with saloon interiors, were used. These were later augmented by semi-open coaches from the Wembley Exhibition railway. The stock was painted dark red, lettered gold or yellow. The Lincolnshire Coast Light Rly, opened in 1960, acquired two of the Ashover Gloucester-built coaches in 1961, mounted them on new bogies and used them in service. Their seats have come from Glasgow, Leeds and Liverpool tramcars.

Type: Two unclassed Saloons. **Owner:** Lincolnshire Coast Light Rly Co Ltd. **Location:** Humberston, Grimsby.

Pullman Car Co
Hastings Pullmans 1926

Pullman cars were introduced to the SECR Charing Cross to Hastings route in 1922. Six new 57ft wooden-bodied cars were built early in 1926 by Metropolitan C&W. These were first-class cars of 8ft 1in only, as the Hastings route has narrow tunnel clearances which dictate the use of narrow stock. In 1932 the cars were converted to composites and were taken out of service during the war in common with other Pullman cars. Three returned to the Hastings service in 1946 rebuilt as buffet cars. The others converted to first-class cars — including Nos 184/5 — were used on Waterloo-Southampton Docks boat trains. In 1958 all six were repainted green and used as buffet cars on the Southampton service, the Hastings trains having been dieselised. In 1960/1 they were transferred to SR stock and numbered S 7872-7. They were withdrawn in 1963/4.

Nos: 184 *Barbara,* 185 *Theodora.* **Type:** Buffet Car. **Owner:** Tenterden Rly Co Ltd. **Location:** Rolvenden.

Southern Railway
Standard Corridor Stock 1927

The adoption of standard corridor stock on the SR was complicated by limited clearances on the South Eastern Division, particularly between Grove Junction (Tunbridge Wells) and Battle. The 8ft 6in-width stock could run anywhere except on the Tunbridge Wells-Hastings line, which had to be provided with stock only 8ft 0¾in wide over the bodywork. Since profile was also a loading gauge problem, all these coaches were slab-sided. Otherwise the vehicles were of standard type on 58ft underframes with 8ft wheelbase bogies and steel-panelled, wood-framed bodies. The first new SR corridor stock appeared for the Thanet services in 1924. From 1927 Pullman gangways and buckeye couplers were used for all main-line steam stock. Most was withdrawn by 1960/1 but some, such as Nos 5153

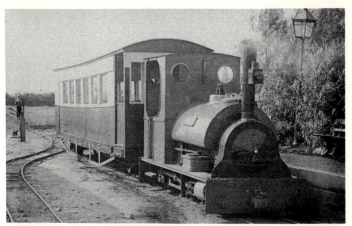

Top: *One of the Ashover Light Railway coaches as preserved.* W. J. K. Davies.
Centre: *One of the Hastings Pullman cars as preserved.* Donald Wilson.
Bottom: *SR No 5618 as preserved.*

5618, survived on workings between Tonbridge-Reading until 1965.

No 5153 (four first-class, three third-class compartments) is one of the 'Folkestone' service stock. From 1929 onwards the corridor side windows were extended up to cantrail level. No 2356, eight compartments, and No 5618 (layout as No 5153) are of this type. No 7798 was originally second class for Folkestone/Dover boat stock, with the unusual layout of two quarterlights flanking the droplight windows on both sides, although of centre gangway internal layout. Nos 1020 (eight compartments) and 7400 (seven compartments) are of the specially narrow Hastings line type. Nos 5153/5618 (KESR Nos 55/56) are in service painted brown and cream. All the others were purchased from service stock and require full restoration.

9ft width stock
No: 5153. **Type:** Composite (1927). **Owner:** Tenterden Rly Co Ltd. **Location:** Rolvenden.
No: 2356. **Type:** Third (1929/30). **Owner:** Bluebell RPS. **Location:** Sheffield Park.
No: 5618. **Type:** Composite (1929/30). **Owner:** Tenterden Rly Co Ltd. **Location:** Rolvenden.
No: 7798. **Type:** Open Second (1929/30). **Owner:** Tenterden Rly Co Ltd. **Location:** Rolvenden.

'Hastings' stock
No: 1020. **Type:** Third (1934). **Owner:** Tenterden Rly Co Ltd. **Location:** Rolvenden.
No: 7400. **Type:** First (1929/30). **Owner:** Tenterden Rly Co Ltd. **Location:** Rolvenden.

London Midland & Scottish Railway
Third-class Sleeping Car   1928

Third-class sleeping car accommodation was introduced on Anglo-Scottish services by both the LMS and LNER in 1928. At first the coaches provided were convertible for day and night use: the upper berths of the four in each compartment were folded back flush with the compartment partition during the day. From the early 1930s both railways introduced purpose-built sleeping cars and fixed berths for third-class passengers.

No 14425 is one of the earlier series of LMS convertible sleeping cars. In general appearance it is similar to the 1923-9 corridor stock. However, this type of sleeping car were 60ft vehicles — as opposed to 57ft — and introduced this length to LMS standard coaches. No 14425 was built as an all-wooden coach, but has since received some

As purchased for preservation. Strathspey Railway.

steel panelling. There are seven compartments and, as built, separate toilets and lavatories at each vestibule end. The bodysides have long windows on the corridor side and three windows to each compartment on the other. No 14425 was renumbered 531 after 1933. In the 1930s fixed berths were fitted. All coaches of this type were withdrawn between 1955 and 1961; some which had been ambulance coaches during 1939-45 had been subsequently converted as buffet cars and lasted longer. The coach now preserved was adapted for service use as DM 395778 and apart from losing the gangways retains many of its original features. It will be restored to LMS condition.

No: 14425. **Type:** Third Sleeping Car. **Owner:** Strathspey Rly. **Location:** Boat of Garten.

Southern Railway
Standard Corridor Stock   1929

The other type of SR corridor stock had tumblehome (rounded) bodysides of 9ft 3in width, although all other dimensional and constructional details were the same as those of the narrower coaches. The outline followed that of the LSW 'Ironclad' stock in most respects. These coaches were barred from working between Tonbridge and Tunbridge Wells and the Hastings line. No 1365, an open third originally seating 56, was later converted to a composite dining car and numbered 7841. The large bodyside windows are of the opening, drop-type. No 1309 is largely similar but outwardly different, with riveted steel panels and fixed windows with sliding ventilators. No 7864 was originally a kitchen/diner converted to a buffet car some time in the mid-1950s. It is now a static refreshment room at Sheffield Park, although still on its bogies. Nos 6575/6686 have two first- and four third-class compartments. Both have the corridor windows carried up to cantrail level. No 6575 has the characteristic side lookouts to the guard's compartment. It was withdrawn from the Exeter area in 1960, with LSW No 320, and these two coaches were the Bluebell's first passenger stock. No 6686, with similar body panelling to No 1309, was equipped with electric train heating and used on the 'Night Ferry' until the mid-1960s. It has now been impressively restored in sage green SR livery. All of this 1929-36 corridor stock had been withdrawn by 1965.

No: 1365. **Type:** Open Third (1929). **Owner:** Bluebell RPS. **Location:** Sheffield Park.

No: 7864. **Type:** Buffet Car (1932). **Owner:** Bluebell RPS. **Location:** Sheffield Park.

No: 6575. **Type:** Composite Brake (1932). **Owner:** Bluebell RPS. **Location:** Sheffield Park.

No: 1309. **Type:** Open Third (1935). **Owner:** Southern Loco Pres Co Ltd. **Location:** Sheffield Park.

No: 6686. **Type:** Composite Brake (1935). **Owner:** Southern Loco Pres Co Ltd. **Location:** Sheffield Park.

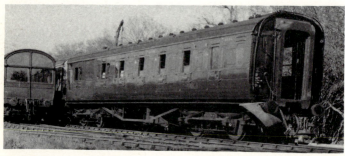

No 6686 as restored. J. Everitt.

No 187 as built — of the same batch as 190.

During 1928-36 the remaining GWR steam railcars were mostly converted to auto-trailers and only 32 new coaches of this type were built. No 167 (one of a batch, Nos 159-70) is a bow-ended 59ft 6in coach, 9ft in width, running on 7ft-wheelbase type bogies. The two saloons on either side of the entrance gangway seat 64. Nos 178/190 are of the longer 62ft 8in length variety seating 73. All these coaches have large bodyside windows with narrow inward-opening ventilators below the cantrail. No 178 has 7ft bogies, No 190 the later 9ft type. The main difference between the three vehicles is that No 190 has flush-panelled bodysides. All of these auto-trailers were withdrawn between 1958 and 1961, a number going to departmental use as in the case of Nos 178/190. No 190 has been restored and used on Didcot open day workings.

No: 167. **Type:** Third Auto-trailer (1929). **Owner:** Dean Forest RPS. **Location:** Parkend.
No: 178. **Type:** Third Auto-trailer (1930). **Owner:** Private. **Location:** Bewdley.
No: 190. **Type:** Third Auto-trailer (1933). **Owner:** Great Western Soc Ltd. **Location:** Didcot.

Underground Group
Tube Stock Motor Car 1929

Unfortunately no early tube cars have been preserved, apart from the City and South London trailers (page 00). Great Northern, Piccadilly & Brompton Rly motor car No 51 of 1906 was put aside for

No 3327 prepares to enter the Science Museum.

preservation in the early 1950s, but subsequently scrapped, apart from a body section retained for display at Syon Park. The motor car on display at the Science Museum is one of the general series known as the 'Standard Stock' built between 1923-34. These cars were all-steel, the motor cars just over 51ft in length with the switchgear mounted above the underframe, which was upswept over the motor bogie. Air-operated doors were fitted. Motor car No 3327, originally No 297, went into service in July 1929 on the Piccadilly Line; it was one of an order placed in 1927 for 306 motor and trailer cars, most of which went to the Northern Line. The motor cars in this order had control equipment supplied by British Thomson-Houston with General Electric Company traction motors. It later went to the Bakerloo Line and then, from March 1939, to the Central Line, until replaced by the '1959 Tube Stock' in 1961. The car has been restored to its latter-day condition. Some of the 1923-34 Tube Stock is still running on the BR Isle of Wight line between Ryde and Shanklin.

No: 3327. **Type:** Unclassed Motor Car. **Owner:** Department of Education & Science. **Location:** The Science Museum, London.

146

Great Western Railway
Dining Cars

Very few restaurant cars — as opposed to buffets — from this period have been preserved so that these three are of particular interest. All are bow-ended 61ft 4½in over the ends and 9ft 3in wide (No 9605 9ft wide); 9ft wheelbase bogies are used. Ten each of the first kitchen and third diners, such as 9615/27, were built to work in pairs, the first/kitchen car being coupled with the kitchen next to the third diner. The total seating capacity was 86: 23 first class and 63 third class. Around 1946, in common with most GWR catering vehicles, all of these pairs were renovated, being refurbished internally with bench type seats and new lighting, and externally by fitting new, larger windows with sliding ventilators. From 1956 nearly all were repainted in chocolate and cream livery for working in the principal WR expresses. Around 1962 they were displaced by standard BR stock and withdrawn. The two preserved coaches rather unusually survived in department stock, No 9615 at Cardiff and No 9627 as an office at Roath Docks, Cardiff. The SVR is putting considerable effort into their restoration and has already re-equipped No 9615 with new catering equipment so that it can provide refreshment facilities on SVR trains.

No 9605 was one of a batch of 10 dining cars built for cross-country services with an end kitchen and two saloons with fixed seating — accommodating 12 first-class and 31 third-class diners. It is hoped that No 9605 will be fully restored and on display at York by mid 1976.

No: 9605. **Type:** Composite Diner (1930). **Owner:** Department of Education & Science (not yet on display).
No: 9615. **Type:** First/Kitchen (1932). **Owner:** Private. **Location:** Bewdley.
No: 9627. **Type:** Third Diner (1932). **Owner:** GW (SVR) Association. **Location:** Bewdley.

Restaurant car pair Nos 9611/21 as built.

Great Western Railway
Saloons

These two fine coaches were built as first-class corridor saloons for private hire. They are bow-ended, 61ft 4in long and 9ft wide with the sides bulging round at the bottomsides. Observation windows are fitted at the body ends. The interior followed the normal layout with saloons at each end, a separate first class compartment and a central kitchen/pantry and lavatory. All saloon interior panelling is in walnut. Both were fitted with new windows in the 1930s and in 1947 lost the separate compartment as well as being modernised internally. Subsequently both were repainted in chocolate and cream livery in BR days. In the early 1960s No 9005, the WR General Manager's saloon, was mounted on the Swindon-developed B4 bogies, the other coach retaining the original 7ft type. No 9004 went to the North Eastern Region as the Civil Engineer's saloon. It was withdrawn in 1972 and restored to GWR livery at the York works of British Rail Engineering Ltd. No 9005 was withdrawn in 1974 and is currently in BR 'Rail Blue' livery, having now exchanged its B4 bogies for the original BR 1951 type.

No: 9004. **Type:** First Saloon. **Owner:** Mr W. H. McAlpine/Flying Scotsman Enterprises. **Location:** Carnforth.
No: 9005. **Type:** First Saloon. **Owner:** Great Western Soc Ltd. **Location:** Didcot.

No 9004 in the 1930s.

Great Western Railway
'Super' Saloons

These coaches are among the most famous built by the GWR and deservedly so in view of their handsome outline and high-class Pull-

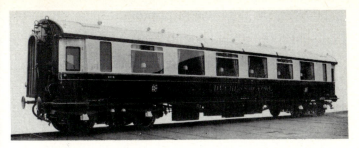

No 9116 as built.

man-type furnishings. Eight were built in 1931/2 for use on the Plymouth-Paddington Ocean Liner expresses, where they replaced the Pullman cars used from 1929-31. The 'Super Saloons' are more handsome in appearance than the contemporary Pullman cars and also wider as the extreme width of the 61ft 4in body is 9ft 7in. As a result, they were subject to some restrictions on route availability and the doorways were recessed at an angle to the bodywork. Standard 9ft bogies were fitted. Before 1939 they carried the names shown above. The interior furnishings are very lush with walnut panelling and single armchair seats on either side of the gangway. There is also a single compartment at one end. In 1935 Nos 9117/8 were rebuilt with a kitchen in place of this compartment. After the war the 'Super Saloons' were generally renovated and the large windows received sliding ventilators. The Plymouth boat services were discontinued from the end of 1962, but the 'Super Saloons' continued on special duties as well as making regular appearances on the Newbury Racecourse specials. Happily, five out of the eight were bought for preservation on withdrawal in 1966/7 so that we can continue to admire these fine examples of GWR coachbuilding. Nos 9112/8 are approved by BR for main-line running.

No: 9111 *King George.* **Type:** First Saloon. **Owner:** Dart Valley Rly Association. **Location:** Buckfastleigh.

No: 9112 *Queen Mary.* **Type:** First Saloon. **Owner:** Great Western Soc Ltd. **Location:** Didcot.

No: 9113 *Prince of Wales.* **Type:** First Saloon. **Owner:** Private. **Location:** Didcot.

No: 9116 *Duchess of York.* **Type:** First Saloon. **Owner:** Dart Valley Rly Co Ltd. **Location:** Buckfastleigh.

No: 9118 *Princess Elizabeth.* **Type:** First Saloon. **Owner:** Great Western Soc Ltd. **Location:** Didcot.

Manchester South Junction & Altrincham Railway Electric Stock Trailers 1931

The Manchester South Junction & Altrincham Railway was a jointly controlled railway — latterly by the LMS and LNER — remaining as such until nationalisation. In 1931 it became the first British passenger line to be electrified at 1500V dc overhead. The multiple-unit stock for the electrified services was of LMS design built by Metropolitan-Cammell. Twenty-two three-car sets were supplied — a motor coach, composite trailer and driving trailer to each set. Although very similar in appearance to contemporary LMS steam and electric suburban stock, the coaches were of an otherwise non-standard 58ft length. The MSJA electric stock was painted in a green livery and numbered in a special series until 1948; thereafter it was renumbered and painted in standard BR liveries. Nearly all of this electric stock remained until the conversion of the Altrincham line to 25kV ac electrification in 1971, when it was withdrawn.

The three surviving coaches were virtually snatched from the scrapyard and are of the same nine-compartment type, of which the centre five compartments were first class. The two coaches at Embsay now run behind steam traction on open days.

No: 114 (29663). **Type:** Composite Trailer. **Owner:** Derby Corporation. **Location:** Butterley.
Nos: 117 (29666), 121 (29670). **Type:** Composite Trailer. **Owner:** Yorkshire Dales Rly Co Ltd. **Location:** Embsay.

MSJA electric set in BR days. P.J. Sharpe.

Pullman Car Co
'Brighton Belle'

The 'Belle' as best remembered, in 1959. J.C. Beckett.

From New Year's Day 1933 three five-car electric multiple-unit Pullman sets entered service on the 'Southern Belle' service between Victoria and Brighton. Two sets were used for the train throughout its life, with one held as a spare. The fifteen all-steel coaches making up the '5-BEL' sets were built by Metropolitan-Cammell in 1932. Each set consisted of a motor brake third at each end (66ft 8¾in long), seating 48 passengers; two first class kitchen cars (66ft long), each seating 20 passengers; and a third parlour car (66ft long), seating 56 passengers. Pullman tradition was maintained by both the characteristic livery and the interior furnishings. From 29 June 1934 the service was renamed 'The Brighton Belle.' Until 1937 the sets were numbered 2051-3, thereafter Nos 3051-3. The 'Belle' did not run between 1939/40 and 1942-6 and during the war set No 3052 was bombed, but rebuilt in 1946/7. The 5-BEL units then settled down to a long period of reliable service, receiving new bogies in 1955 to cure their rough riding (although the new bogies were little better!). From 1969 the units were 'jazzed up' by being repainted in BR blue and grey livery and they received new upholstery. The first-class cars then lost their names. On 30 April 1972 the 'Belle' ran for the last time and soon after the cars were put up for sale.

Unit 3051 (2051)
Name: Car No 88. **Type:** Third Parlour Motor Car. **Owner:** Private. **Location:** Chappel.

Name: *Hazel* (279). **Type:** First Kitchen. **Owner:** "Black Bull" Inn. **Location:** Moulton, Yorkshire.

Name: *Doris* (282). **Type:** First Kitchen. **Owner:** Private. **Location:** Finsbury Park, London.

Name: Car No 86. **Type:** Third Parlour. **Owner:** Private. **Location:** (in store).

Name: Car No 89. **Type:** Third Parlour Motor Car. **Owner:** "Little Mill" Inn. **Location:** Rowarth, Stockport.

Unit 3052 (2052)

Name: Car No 90. **Type:** Third Parlour Motor Car. **Owner:** British Railways Board (in store).

Name: *Audrey* (280). **Type:** First Kitchen. **Owner:** South Eastern Steam Centre. **Location:** Ashford (Kent).

Name: *Vera* (284). **Type:** First Kitchen. **Owner:** Private. **Location:** Westleton (Suffolk).

Name: Car No 87. **Type:** Third Parlour. **Owner:** North Norfolk Railway. **Location:** Sheringham.

Name: Car No 91. **Type:** Third Parlour Motor Car. **Owner:** North Norfolk Railway. **Location:** Sheringham.

Unit 3053 (2053)

Name: Car No 92. **Type:** Third Parlour Motor Car. **Owner:** Allied Breweries Ltd. **Location:** (in store).

Name: *Gwen* (281). **Type:** First Kitchen. **Owner:** "Horseless Carriage" Inn. **Location:** Chingford.

Name: *Mona* (283). **Type:** First Kitchen. **Owner:** "Brighton Belle" Inn. **Location:** Winsford (Cheshire).

Name: Car No 85. **Type:** Third Parlour. **Owner:** Private. **Location:** (in store). Breweries Ltd. **Location:** (in store).

Name: Car No 93. **Type:** Third Parlour (Motor Car). **Owner:** Allied Breweries Ltd. **Location:** Manningtree (in store).

Southern Railway
6-PUL/6-PAN Electric Stock   1932

The electric multiple-unit stock for the Brighton main-line electrification in 1933 consisted of 23 six-car corridor units which included a Pullman car. The motor cars at each end were centre-gangway coaches and the trailer cars of side corridor layout. The motor cars and Pullmans were of all-steel construction, but the other trailer cars were steel-panelled on wooden body frames. The four motor bogies of the two motor cars could develop 1800hp and the maximum speed was 75mph. Further six-car sets were built in 1935 for the Hastings service. All were withdrawn from the Brighton/Hastings workings by

A motor car of set 3001 in BR days. P.J. Sharpe.

1965, but some sets in reformed state survived until 1968.

The trailer composite of this collection has five first-class and three third-class compartments. It was marshalled in a reformed 4-COR set (*see page 168*) and was in service until 1972. The two Pullman cars were built by Metro-Cammell and provided 12 first and 16 third-class seats. The overall length is 66ft and the body width 9ft. *Bertha* was originally in set No 3001, *Ruth* in set No 3042. *Ruth* has now been repainted in its original livery. The other coaches await restoration.

No: 11773. **Type:** Trailer Composite. **Owner:** Private. **Location:** Ashford (Kent).

Pullman Cars
Name: *Ruth.* **Type:** Composite Kitchen. **Owner:** 6000 Locomotive Association. **Location:** Hereford.
Name: *Bertha.* **Type:** Composite Kitchen. **Owner:** South Eastern Steam Centre. **Location:** Ashford (Kent).

Great Western Railway
Standard Corridor Stock   1933

Having built bow-ended corridor stock on 60ft underframes from 1929 to 1933 the GWR reverted to a 57ft flat-ended design between

Brake Third No 5878 as built.

1933-5. This type continued with the traditional layout of an exterior door to each compartment and four doors to the corridor side. The pressed-steel 9ft wheelbase bogie was introduced with this design. These coaches were in service until 1964.

Nos 5787/5883 are four-compartment brake thirds, virtually identical except for the layout of the luggage compartment doors. No 5787 has been fully restored and is in service, while No 5883 is in process of renovation.

No 5952, an eight-compartment coach, was purchased when withdrawn from traffic. It too has been fully restored and 'plated' for 'Return to Steam' specials over BR lines.

Nos 6912/3 have two first and four third-class compartments. Both were purchased from Radyr in 1974.

No 111 is a full-length van.

No: 5787. **Type:** Third Brake (1933). **Owner:** Private. **Location:** Didcot.
No: 5883. **Type:** Third Brake (1934). **Owner:** Great Western (SVR) Association. **Location:** Bewdley.
Nos: 6912, 6913. **Type:** Brake Composite (1934). **Owner:** Great Western (SVR) Association. **Location:** Bewdley.
No: 5952. **Type:** Third (1935). **Owner:** Great Western Soc Ltd. **Location:** Didcot.
No: 111. **Type:** Passenger Brake Van (1934). **Owner:** Great Western Soc. **Location:** Didcot.

London Midland & Scottish Railway
Standard Corridor Stock   1934

The LMS was the first of the 'Big Four' to standardise a modern corridor stock design which had the end vestibule layout and large windows with sliding ventilators for both bodysides. These coaches were steel-panelled with wooden body framing. The layout and construction were highly standardised. The composites and brake composites were 60ft long, all other ordinary corridor coaches 57ft. All were 9ft 3in wide. A standard 9ft wheelbase bogie was used throughout. The first of these standard corridor coaches appeared in 1933 and they were built with very few major changes into post-war days (*see also page 179*).

No 7511 is an exception to the above as it was built on a 65ft underframe. In original condition it seated 42 in two dining saloons. The interior was spacious and finished to a high standard with Pullman-type features. In the early 1960s it was gutted and converted to an exhibition coach. As yet it has not been restored, but is in use for volunteer sleeping accommodation. No 5987 is a four-compartment brake third and was purchased when withdrawn and is intact. Full restoration is due to start shortly. No 9388, an open third, originally seated 60 in two saloons. It was converted to an exhibition coach. From 1974 it has been used as a 'bar coach' in SVR trains. and will soon be further restored and painted.

No: 7511. **Type:** Restaurant Open First (1934). **Owner:** Severn Valley Rly Co Ltd. **Location:** Bewdley.
No: 9388. **Type:** Open Third (1936). **Owner:** Severn Valley Rly Co Ltd. **Location:** Bewdley.
No: 5987. **Type:** Side-corridor Third Brake (1937). **Owner:** Princess Elizabeth Loco Soc. **Location:** Ashchurch.

Open Third No 9443 as built.

LMS (Northern Counties Committee)
'North Atlantic' Third Brake (5ft 3in gauge)   1934

The first coach in the 'Portrush Flyer' in 1974. C. P. Friel.

In 1934 the NCC improved its main line from Belfast to Londonderry with the construction of the Greenisland loop enabling
through running in place of the previous reversal at Greenisland.
From June 1934 a new timetable was introduced on the main line
and included a fast, light train between Belfast and Portrush named
the 'North Atlantic Express.' New stock was built in 1934/5 for this
service which was of side-corridor layout — a brake third, two thirds
and a composite. These coaches were 57ft in length and very similar
to the one-window-per-compartment corridor stock built by the LMS
in the early 1930s. Toplight ventilators under the cantrails were a
particular feature of the 'North Atlantic Stock.' Nos 91-4 remained
as locomotive-hauled stock throughout their lives and three were still
in service in 1973. No 91 — renumbered 472 in 1959 — was purchased by the RPSI as part of the bulk purchase in 1975 referred to
on page 124. It had already been used on a number of occasions on
RPSI excursions.

No: 91 (472). **Type:** Third Brake. **Owner:** Railway Preservation Soc of
Ireland. **Location:** Whitehead.

Southern Railway
Nondescript 'Boat' Brakes   1934

These coaches were built as second-class vehicles for service in SR
Continental boat trains on the South Eastern Division. The body

length is 60ft and the width 9ft and the coaches are of standard SR corridor stock specification (*see page 140*). The interior with centre gangway consists of one two-bay and one four-bay saloon, seating 36 in all in a two and one-a-side layout. Apart from four coaches converted to invalid saloons in 1960, all remained in use on boat trains until 1961/2. Some were then used on Tonbridge-Brighton/Reading trains until 1964, such as Nos 4432/43. No 4441 was converted to service stock and awaits full restoration. The other two have been repainted in the Tenterden Railway's chocolate and cream livery and are in use, numbered 53/54.

Nos: 4432, 4443. **Type:** Brake Second Open. **Owner:** Tenterden Rly Co Ltd. **Location:** Rolvenden.
No: 4441. **Type:** Brake Second Open. **Owner:** Bluebell RPS. **Location:** Sheffield Park.

Either No 4432/43 as restored. Brian Morrison.

Great Western Railways
Buffet Car
1934

During the 1930s all four railways introduced buffet cars, but the two GWR cars of 1934 were the most unusual. Although carrying the 'Buffet Car' designation on the bodysides they were also referred to as 'Quick Lunch Bar Cars.' The interior resembled a cocktail bar: there was a pantry at one end and the rest of the interior was made up of a continuous bar counter faced by twelve stand-up bar stools.

An interior (of Nos 9631/2) for a change.

Showcases and tea urns were placed on the bar counter. The windows behind the bar were of frosted glass, while on the other side there were large windows high placed on the bodyside. The coaches were built on the 57ft underframe and the flat-ended body was of similar profile and 9ft width to the 1933-5 corridor stock. Nos 9631/2 were among the first coaches to carry the GWR monogram. They first appeared on the Paddington-Worcester and Paddington-Bristol lines, one later being on the 'Bristolian.'

No 9632 was withdrawn in 1963, but the other car was restored by Swindon Works in 1961/2 — internally as well as externally — to original condition. It was on display out of doors at Clapham Museum from 1962-73, but was then transferred on loan to the Severn Valley Railway for a period of eight years.

No: 9631. **Type:** Buffet Car. **Owner:** Department of Education & Science. **Location:** Bewdley (on loan).

Great Western Railway
Diesel Railcars   1934

Early experiments with internal combustion-engined railcars on British railways were not very successful. The GWR was the only one of the main-line railways to evolve a really useful design. The first car, No 1, appeared in December 1933 and introduced a completely new outline with its fully streamlined body. The earlier cars, Nos 1-18 — all third class — were built by outside contractors and powered by AEC bus-type diesel engines. Some were used on express services and were provided with buffets, while others were built for branch-line work. No 4, a double-ended car with bodywork by Park Royal, was outshopped in September 1934 for use on the Birmingham-Cardiff service. Its overall length is 60ft and it is 9ft wide. The interior seats 44 in two saloons with a small end buffet. It was in service until 1959 and was then fully restored at Swindon Works in the following year. Although preserved in the first place by BR it was transferred on loan to the GW Society in the late 1960s by Swindon Corporation (who had taken over some GWR relics).

Nos 20/22, both double-ended cars, were branch-line railcars built by the GWR — again with AEC engines. These Swindon-built cars had a more angular appearance with slab-sided bodies. The interiors have seating and furnishings similar to contemporary excursion stock, rather than the bus-type designs of BR railcars. Both Nos 20/22 were in service until 1962 and were among the last of their kind in traffic. No 22, beautifully restored, is in regular use for the 'off-peak' SVR trains and the intention is that No 20 should be similarly employed down in Kent.

No: 4. Type: Diesel Railcar (1934). **Owner:** Swindon Corporation. **Location:** Didcot (on loan).

No: 20. Type: Diesel Railcar (1940). **Owner:** Tenterden Rly Co Ltd. **Location:** Rolvenden.

No: 22. Type: Diesel Railcar (1941). **Owner:** GW Preservations Ltd. **Location:** Bewdley.

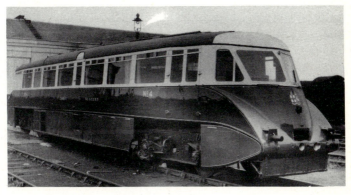

No 4 as restored. G. Wheeler.

London & North Eastern Railway
Sleeping Cars   1935

The first really modern sleeping cars appeared in 1930, when efficient pressure-ventilation systems had been developed. In the same year the LNER also produced the innovation of a shower compartment in one of its cars. Thought was given at the same time to the development of attractive and serviceable interior finishes and fittings. A 65ft underframe was used for two first-class sleepers in 1932; this length enabled ten berths and a shower bath to be provided. Further coaches of this type were built in 1934/5, including No 1211. These vehicles had teak-panelled bodies with the characteristic sloping roof ends. All went into service on the principal East Coast Anglo-Scottish sleeping car trains. The two centre berths of the coaches could be converted to form a single day/sleeper compartment if so wished. Most of these 65ft sleepers — 66ft 6in over the ends and 9ft 3in wide — lasted into the 1960s. No 1211, as the photograph shows, survived long enough to be painted in BR blue/grey livery in 1968. It was withdrawn as the last wooden-bodied sleeping car in early 1972. The Strathspey Railway Association is now restoring it to LNER varnished teak livery.

A similar first-class sleeper, No 1592, which had been used in General Eisenhower's European Forces HQ train during the Second World War, was bought for the American National Railroad Museum at Green Bay, Wisconsin, and exported in 1969.

No: 1211. **Type:** First Sleeper. **Owner:** Strathspey Rly. **Location:** Boat of Garten.

No 1211 in BR livery 1968. D. L. Percival.

Great Western Railway
'Centenary' Restaurant Car

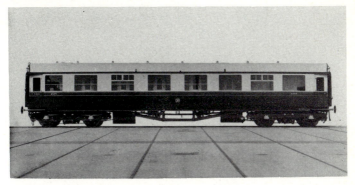

'Centenary' brake composite No 6659.

In the year of its centenary the GWR built two sets of coaches of an entirely new design for the Cornish Riviera Limited. The 'Centenary' stock, as it became known, took advantage of the old broad-gauge loading gauge and the coaches were built to the maximum width of 9ft 7in, the bodies being bow-ended and 61ft 4½in long over the ends. As with the 'Super Saloons' (*page 148*) the doorways were recessed. For the first time since the 'Dreadnought' stock of 1905 passenger access was via end vestibules only so that large compartment side windows could be used. Enough coaches were built to make up two ten-coach sets together with spares for strengthening at holiday periods. The interior decor was restrained with polished wood panelling. The restaurant car pairs consisted of a first/kitchen and a third-class saloon. The other coaches built were brake thirds, thirds, composites and brake composites. This stock was used on the 'Cornish Riviera' from 1935 until about 1941. Thereafter it seldom worked together in sets and was withdrawn during 1962-4.

Only one vehicle has survived — restaurant car No 9635, which was purchased on withdrawal in 1963. It was modernised in early post-war days with new interior furnishings, including low-backed bench-type seats. Full restoration was started in 1972.

No: 9635. **Type:** First Kitchen/Restaurant. **Owner:** Dowty RPS. **Location:** Ashchurch.

Southern Railway
2-BIL Electric Stock

2-BIL unit in service in the 1960s. J.H. Bird.

These two-coach electric multiple-units were built for semi-fast services and were used on the SR Central Division and Waterloo-Reading/Alton/Portsmouth lines. The sets consisted of side-corridor vehicles: a third-class motor car and a composite driving trailer which provided 24 first-class and 84 third-class seats. The coaches were similar to contemporary steam stock in construction and facilities. Originally numbered in the 18xx/19xx series, they were renumbered 2001-2152 in 1937. General withdrawal began in 1969, but the last sets, in use on South Coast stopping services, continued until July 1971.

No: 2090. **Type:** 2-BIL Set. **Owner:** Department of Education & Science. **Location:** In store.

Great Southern Railway (Ireland)
Corridor/Open Stock — (5ft 3in gauge)

The Great Southern Railway, formed in 1925 from the major railways in Southern Ireland, operated under difficult economic conditions during the 1920s and 1930s. Few coaches were built until 1934, when production started of modern steel-panelled types at

Inchicore Works. This stock was withdrawn from 1973 onwards with the arrival of new air-conditioned coaches.

Nos 1327/8 are seven-compartment thirds which seat 56. They are 59ft long and 9ft wide. The bodies are very similar to LMS corridor stock of the period, having large bodyside windows with sliding ventilators and end entrance vestibules. No 1335 is similar but of 9ft 6in body width, with bulging bodysides which taper inwards to the cantrails. The accommodation and other dimensions are the same as for Nos 1327/8. No 1333 was one of a set of coaches built for the Dublin to Bray suburban service. Originally it was non-corridor, but gangway connections were later put in, reducing the seating capacity at the time that this stock was transferred to main-line services. The coach is 59ft and 9ft 3in wide, seating 72 passengers in nine bays; it has neither lavatories nor end vestibules.

All four coaches were purchased from CIE in mid-1973 and have been used on RPSI excursions. No 1333 has been repainted in the RPSI livery of maroon and brown. The coaches are known as the 'Bredins' after the CME of the GSR responsible for their design.

Nos: 1327, 1328. **Type:** Third (1935). **Owner:** Private. **Location:** Whitehead.
No: 1333. **Type:** Open Third (1936). **Owner:** Private (RPSI Members). **Location:** Whitehead.
No: 1335. **Type:** Third (1937). **Owner:** Private (RPSI Members). **Location:** Whitehead.

No 1333 as purchased for preservation. C.P. Friel.

London & North Eastern Railway
Bogie Brake Vans
1936

Among the first elliptical-roofed teak-bodied coaches designed by H. N. (later Sir Nigel) Gresley were bogie brake vans which appeared in 1906. Similar vehicles, 61ft 6in over the ends, were built by the LNER from 1923-43. From 1936 some were built with steel body panelling. A number were equipped with hinged shelves for carrying baskets containing racing pigeons. Some number of these pre-war LNER bogie vans were still in traffic in 1975.

Nos 4236, 4247, 4271 and 70759 are teak-bodied vehicles, No 4247 being used as a mobile sales-publicity shop by the KWVR. Nos 4052 and 4149 are steel-panelled; all have 8ft Gresley single-bolster bogies.

The standard post-war type was 63ft over the ends, steel panelled and with the familiar small lights under the cantrails. No 100 is in use on works trains and No 110 will be used for mobile exhibitions.

No: 4149 (70361). **Type:** Bogie Brake Van (1936). **Owner:** The Lakeside Rly Soc. **Location:** Haverthwaite.

No: 4236 (70459). **Type:** Pigeon Bogie Brake Van (1938). **Owner:** Private. **Location:** Bewdley.

No: 4247 (70470). **Type:** Pigeon Bogie Brake Van (1938). **Owner:** Keighley & Worth Valley RPS. **Location:** Haworth.

No: 4271 (70494). **Type:** Pigeon Bogie Brake Van (1940). **Owner:** Scottish RPS. **Location:** Falkirk.

No: 4052 (70442). **Type:** Pigeon Bogie Brake Van (1941). **Owner:** Keighley & Worth Valley RPS. **Location:** Haworth.

No: 70759. **Type:** Pigeon Bogie Brake Van (1943). **Owner:** David & Charles Holdings Ltd. **Location:** Newton Abbot.

No: 110. **Type:** Bogie Brake Van (1947). **Owner:** North York Moors Rly. **Location:** Grosmont.

No: 100. **Type:** Bogie Brake Van (1948). **Owner:** Keighley & Worth Valley RPS. **Location:** Haworth.

Bogie Brake Van No 4028.

Buffet car No 9135 in service 1973. R.E. Ruffell.

The LNER first introduced buffet cars in 1932, converted from GNR open thirds. The standard design appeared in 1933. This was a bow-ended teak-bodied 61ft 6in vehicle with kitchen, bar counter and a saloon of four bays seating 24. The earliest examples had gas cooking, but from 1935 electric equipment was fitted. The interiors had rexine wall coverings and chromium-plated fittings. In all, 25 were built, being used on a variety of scheduled services as well as on excursion work. Most were refitted internally and equipped with propane gas cooking equipment by BR from 1959. Some then went to other Regions and five were in service into 1975. Five have been preserved. Of these, No 641 has been restored to LNER varnished teak livery and is used for specials on the Dart Valley Railway.

No: 24082 (9118). **Type:** Buffet Car (1936). **Owner:** Quainton Rly Soc Ltd. **Location:** Quainton Road.

No: 24278 (9122). **Type:** Buffet Car (1937). **Owner:** 7597 Group. **Location:** In store.

No: 24280 (9124). **Type:** Buffet Car (1937). **Owner:** Main Line Steam Trust Ltd. **Location:** Loughborough.

No: 641 (9129). **Type:** Buffet Car (1937). **Owner:** Dart Valley Lt Rly. **Location:** Buckfastleigh.

No: 649 (9134). **Type:** Buffet Car (1937). **Owner:** North Yorkshire Moors Rly. **Location:** Grosmont.

London & North Eastern Railway
TPO Sorting Van   1937

The LNER TPO vans were teak-bodied vehicles on 60ft underframes. The bodies were flat-ended without the familiar sloping ends to the roofs. Light-type 8ft 6in bogies were used. No 2441 was built as a sorting van for the London-Newcastle/Edinburgh services. It was originally equipped with TPO collection and delivery equipment. The body is 8ft 6¾in wide. The interior consists of the usual sorting racks, cooking equipment and lavatory. As No 70294 it was one of the last wooden-bodied LNER TPO vans in service and was withdrawn in 1973. No 6777 is 51ft 1½in in length and 9ft wide and was built at Dukinfield in 1931 as a pigeon van for the GE section. At a later date it was converted as a tender (storage van) for TPO services and has GER-design 8ft wheelbase bogies.

The 7597 Fund has also purchased lineside mail pick-up and collection equipment from the Post Office and eventually hopes to restore the sorting van, No 2441, to full working condition.

No: 2441 (70294). **Type:** TPO Sorting Van. **Owner:** 7597 Fund. **Location:** In store.
No: 6777 (70268). **Type:** TPO Tender. **Owner:** 7597 Fund. **Location:** Chappel.

No 70294 as preserved. 7597 Fund.

Great Western Railway
Excursion Stock   1937

The GWR did not build any centre-gangway open stock for general service, unlike the others of the 'Big Four.' The open coaches built

No 1289 as restored — and with full trimmings. J. Hosegood.

from 1935-40 were kept for most of their lives in sets for excursion use. The first set appeared in 1935 and others, to a different design, in 1937/8/40. The coaches in the later sets were 60ft long and of 9ft width (the two 1937 sets); those built in 1938/40 were of slightly different dimensions. All ran on the standard 9ft bogies. The bodies were of modern appearance with large windows with sliding ventilators. The interiors were in contemporary style with square outlines to seat frames, partition doorways and lightshades. During the war a number of the coaches went to the Continent for use in ambulance trains. All were withdrawn between 1962-6.

The three preserved coaches are from the 1937 sets. Each has two saloons with end and centre vestibules and seats 64 passengers. All three have been restored and are used regularly, the Dart Valley pair on the Buckfastleigh line and No 1289, 'plated' for steam and diesel excursions, running over BR lines.

Nos: 1285, 1295. **Type:** Open Third. **Owner:** Dart Valley Lt Rly. **Location:** Buckfastleigh.
No: 1289. **Type:** Open Third. **Owner:** Great Western Soc Ltd. **Location:** Didcot.

Set 3142 in service 1972. J. Scrace.

The SR electrified the Portsmouth 'Direct' line in 1937 and the Mid-Sussex line in 1938. For express services over both routes three types of four-coach corridor multiple-units were built: 55 (later 58) 4-COR without catering facilities, 19 4-RES with a restaurant and kitchen, and 13 4-BUF (usually on the Mid-Sussex route) with a buffet car. Corridor connections at each end of the units allowed through communication. The motor cars were of centre gangway type seating 52 and the trailers of side-corridor layout — of which there was one composite and one third in the 4-COR sets. The coaches were of wooden-framed, steel-panelled construction.

The four motor coaches are all from 4-COR units; the trailer third has eight and a half compartments seating 68, and the composite five first-class and three third-class, seating 54. The buffet car is notable for its Moorish interior decor. It contains a bar, saloon area and kitchen, all of a very distinctive 1930s flavour.

Set no 3142 has complete electrical equipment and work is now in progress to restore it to its SR sage green livery of 1937/8. The trailer Nos 11773 and 11825 are also being renovated.

The whole series of units worked on the Portsmouth/South Coast express services into 1970. Thereafter the 4-COR sets were transferred to stopping services and were all withdrawn during 1972.

No: 11179 (ex set 3131). **Type:** Motor Brake Third. **Owner:** Department of Education & Science. **Location:** In store.

No: 11187 (set 3135). **Type:** Motor Brake Third. **Owner:** Private. **Location:** (in store).

Nos: 11161 (set 3142), 11201 (set 3142). **Type:** Motor Brake Third. **Owner:** Southern Electric Group. **Location: Ashford (Kent).**

No: 11825 (ex set 3135). **Type:** Trailer Composite. **Owner:** Private. **Location:** Ashford (Kent).

No: 10096 (ex set 3142). **Type:** Trailer Third. **Owner:** Southern Electric Group. **Location:** Ashford (Kent).

No: 12529 (ex set 3084). **Type:** Buffet Car (1938). **Owner:** Department of Education & Science. **Location:** In store.

London & North Eastern Railway
Coronation Observation Saloons   1937

The story of the LNER's streamlined express trains is well known. Unfortunately, out of all the fine vehicles built for the 'Silver Jubilee', 'Coronation' and 'West Riding Limited' from 1935-8 only the two 'Coronation' observation saloons remain. The other coaches were all scrapped by 1964. The 'Coronation' ran as a nine-coach train — four articulated twins and the observation saloon — although the latter was included in the train during the summer months only. The observation saloons, like the other coaches, had a wooden-framed body with steel panelling and stainless steel trim. These saloons are

No 1729 after rebuilding, 1959. K.R. Pirt.

51ft 9in in length over the body and 9ft 2¼ in wide. One end slopes
downwards in matching shape to the front end of the 'A4' streamlined Pacific locomotives. The interior seated 16 in large armchairs.
At the train end of the coach was a compartment for letter mails.
The 'Coronation' stock was stored from 1940 and all except the
observation saloons reappeared from 1948. The observation saloons
were used occasionally until 1956, when No 1729 commenced regular
summer service on the West Highland line, to be joined later by No
1719. Both were rebuilt in 1959 with a more angular observation end
which provided better viewing. Both were used on the West Highland route until the end of summer 1967. No 1729 was formerly on
the KWVR. No 1719 is used on the Lochty Railway.

No: 1719. **Type:** Observation Saloon. **Owner:** Lochty Private Rly. **Location:**
Lochty (Fife).
No: 1729. **Type:** Observation Saloon. **Owner:** The Gresley Soc. **Location:**
Ashford (Kent).

Great Western Railway
Standard Corridor Stock   1938

Third No 1080 as built.

The GWR moved forward to a modern layout for corridor stock in
1936: entry through end vestibules only; large windows on both
corridor and compartment sides, and more attractive interiors.
These coaches were of differing lengths: some 58ft, others 59ft or
60ft. As from 1938 a modified profile and 8ft 11in width was adopted
to allow running over most of the lines of the other Big Four. The
bodies were wooden-framed with steel body and roof panelling. The

standard pressed-steel 9ft wheelbase bogie was used. The last of the type appeared in 1941. All were withdrawn by 1966. Some were then used in the Swindon works test train until 1969. This was lucky for the GW Society and SVR, as an opportunity arose to purchase nine intact coaches which have been comparatively easy to restore for preservation. Most have now been brought back to GWR livery. Nos 536, 1111, 7313/71 are approved and 'plated' by BR for main-line running.

The thirds are all of the eight-compartment layout. The brake composites have two first- and four third-class compartments; the composites have four first- and three third-class compartments. No 6705, a brake composite of 1938, was restored to GWR livery at Swindon in 1967 and exported to Steamtown, Virginia, USA.

No: 6562. **Type:** Brake Composite (1938). **Owner:** Private. **Location:** Bewdley.
No: 1086. **Type:** Third (1938). **Owner:** Private. **Location:** Bewdley.
No: 1087. **Type:** Third (1938). **Owner:** Great Western (SVR) Assoc. **Location:** Bewdley.
No: 1111. **Type:** Third (1938). **Owner:** Great Western Soc Ltd. **Location:** Didcot.
No: 1116. **Type:** Third (1938). **Owner:** Private. **Location:** Bewdley.
No: 1146. **Type:** Third (1938). **Owner:** GWR 813 Preservation Fund. **Location:** Bewdley.
No: 7313. **Type:** Composite (1939). **Owner:** Great Western Soc Ltd. **Location:** Didcot.
No: 536. **Type:** Third (1940). **Owner:** Great Western Soc Ltd. **Location:** Didcot.
No: 7362. **Type:** Brake Composite (1940). **Owner:** Great Western Soc Ltd. **Location:** Didcot.
No: 7284. **Type:** Composite (1941). **Owner:** Private. **Location:** Bewdley.
No: 7285. **Type:** Composite (1941). **Owner:** Private. **Location:** Didcot.
No: 7371. **Type:** Brake Composite (1941). **Owner:** Great Western Soc Ltd. **Location:** Didcot.

London Midland & Scottish Railway
Non-Corridor Third   1938

Nearly all non-corridor stock built by the LMS was on 57ft underframes. Apart from changes in body panelling and constructional methods the basic layout of LMS non-corridor coaches stayed the same from Grouping until the last was built in 1951. The flush steel-panelled design with steel roof panels first appeared in 1933. No 12066 is of this type and has nine compartments seating 108 passengers. The body is 57ft long and 9ft 3in wide, with standard 9ft bogies. It was one of the last survivors; all ex-LMS non-corridor

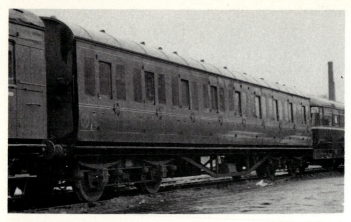

As purchased for preservation. Ian G. Holt.

stock was withdrawn during 1967. It is now painted in a maroon livery and is KWVR No 6.

No: 12066. **Type:** Third. **Owner:** Keighley & Worth Valley RPS. **Location:** Haworth.

Great Northern Railway (Ireland)
Dining Car — (5ft 3in gauge) 1938

The GNR had a good record for modern coaching stock practice and this coach is an example of the steel-panelled stock of the late 1930s. It is steel-panelled on a wooden frame with a wood and canvas roof. The body length is 58ft and the width 9ft 6in. As originally built, it was a composite restaurant car with first-class and second-class saloons and an end kitchen. As such it went into service on the principal Belfast-Dublin trains. In the late 1950s it was converted for use in diesel multiple-unit sets of the high-powered BUT-engined type and worked on the Belfast-based 'Enterprise' from April 1958, by then as UTA No 552. At a later date the fixed seating was replaced by loose chairs and the interior modernised with fluorescent lighting and plastic panelling. During early 1974 RPSI members put in a lot of work on this vehicle; new ceiling and saloon panelling were fitted, the roof re-felted and the kitchen rebuilt to provide a counter and bar area. As such it has been used during 1974 on RPSI excursions and will later be restored to GNR livery.

No: 88. **Type:** Kitchen/Dining Car. **Owner:** Private (RPSI members). **Location:** Whitehead.

After preservation. but still in NIR two-tone livery. C.P. Friel.

London & North Eastern Railway
'Flying Scotsman' Buffet Car   1938

Thirty coaches were built in 1938 for the two 'Flying Scotsman' sets. These were of the standard LNER corridor stock specification with teak-panelled bodies. The catering facilities were provided by an articulated restaurant car triplet and a buffet car. All coaches in the trains were fitted with pressure ventilation. The buffet lounge cars, Nos 1852/3, were 66ft 6in over the ends and 9ft 3in wide. The interiors were unique in that the lounge area was partitioned off by a glass screen. At one end was a buffet/bar area, at the other a ladies retiring room. In 1959 both coaches were rebuilt as conventional buffet cars and equipped with gas cooking equipment (originally they were all-electric). No 1852 was withdrawn in 1965 for preserva-

No 1852 at Ashford.

tion and until 1974 was on the KWVR. It is at present painted in a maroon livery.

No: 1852. **Type:** Buffet Car. **Owner:** The Gresley Soc. **Location:** Ashford (Kent).

London & North Eastern Railway
Restaurant Buffet Car   1939

Two buffet-restaurant cars were built at Dukinfield works in 1939 for use between York and Swindon on the Aberdeen-Penzance service. These had all-electric cooking equipment and the interior was arranged in separate buffet and restaurant areas. The 61ft 6in bodies were of the standard teak-panelled type, 9ft 3in wide. Light-type 8ft 6in compound-bolster bogies were fitted. No 9195 was rebuilt by BR with a modernised interior and gas cooking; in this condition the interior seats 24. During the 1960s it was used on various duties such as the 'Cambridge Buffet Expresses,' and in 1972 appeared on the Harwich-Manchester boat train. It was with-

As restored. T. Boustead.

drawn in 1973 and has now been beautifully restored externally to its original finish; the interior remains as modernised by BR.

No: 24287 (9195). **Type:** Restaurant Buffet Car. **Owner:** Mr W. H. McAlpine/ Flying Scotsman Enterprises. **Location:** Carnforth.

London Midland & Scottish Railway
PO Sorting Van   1939

This vehicle was built as a standard PO sorting van with bag pick-up and collection net apparatus. It is steel-panelled and is one of the 60 ft underframe type built 1930-54, but does not have a staff lavatory. The body width (over the panels) is 8ft 8in. It was originally used for the West Coast postal train from London-Glasgow/Aberdeen. In 1968 it was transferred to the Manchester-Glasgow service, lost its 'lineside' apparatus and was painted in BR blue/grey livery. It was withdrawn in 1973. No 30272, although of later construction, is very similar to 30225 but does have a lavatory.

The lineside collection and delivery of mails from TPOs was dis-

No 30225. As acquired for preservation. J. B. Radford.

continued in 1971. It is good to know that a set of lineside equipment has been acquired so that this interesting aspect of railway working will one day be demonstrated using No 30225, which is now being restored to LMS livery.

No: 30225. **Type:** PO Sorting Van (1939). **Owner:** Derby Corporation. **Location:** Butterley.
No: 30272. **Type:** PO Sorting Van (1950). **Owner:** Department of Education & Science (in store).

Great Western Railway
Special Saloons   1940

These two special saloons were ordered in 1938 and completed in 1940. They were built on a 60ft underframe and given six-wheel bogies. The steel-panelled bodies were 60ft 11½in long and 8ft 11in wide. The interior was divided into a dining saloon with fixed seating, kitchen, a coupé saloon and a day saloon which had armchairs and small tables.

The original intention was that they should be used for daytime journeys by VIPs, but of course they went straight into wartime military service. During the war Nos 9001/2 were used by Sir Winston Churchill and General Eisenhower as well as by members of the Royal Family. In 1944 both were fitted out with comprehensive radio equipment. The original interior furnishings had necessarily been rather spartan, but in 1953 the interiors were repanelled and the

No 9002 as built.

dining saloons had the fixed seating replaced by a central table and loose chairs. As from 1963 Nos 9001/2 were advertised for private hire, particularly for business delegations. Both were used for the last time in 1968 and were sold out of service to become VIP saloons for their new owners.

No: 9001. **Type:** First Saloon. **Owner:** Private. **Location:** Bewdley.
No: 9002. **Type:** First Saloon. **Owner:** Great Western Soc Ltd. **Location:** Didcot.

Great Northern Railway (Ireland)
Open/Corridor Stock — (5ft 3in gauge)   1941

The GNR built modern steel panelled stock from the mid-1930s with large bodyside windows and sliding ventilators — similar to contemporary stock in Britain. Later examples had bodyside panelling of a processed wood material. The open thirds were notable for having four doors to each bodyside. A 10ft wheelbase bogie was used. In 1947/8 the GNR introduced its 'Enterprise' express workings between Belfast and Dublin. For these trains new first and brake first corridors 60ft in length were built, but the other vehicles were existing stock. The interiors were panelled in mahogany and rexine. In 1950/1 the GNR introduced diesel multiple-unit stock and converted some locomotive-hauled coaches to work with them. From 1957 further power cars were put into service and these too were worked with steam stock conversions. Many of the 1935-50 period

177

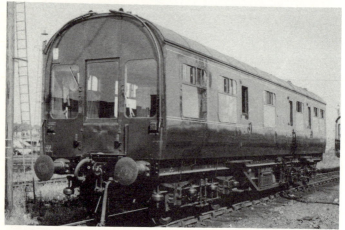

Top: *GNR (Ireland) Nos 227, 176, 175 awaiting delivery to the RPSI.* C.P. Friel.
Centre: *LMSR 1940-7 saloon in service 1969.* D.A. Idle.
Bottom: *Brake Third No 27001 in BR blue/grey livery.*

coaches remained in traffic into the 1960s and some into the 1970s. Six diesel railcars (converted coaches) were purchased by the RPSI as part of its 1975 bulk purchase.

Nos 98/176 are open third coaches, 58ft in length, seating 72. No 98 was converted for railcar use in 1951, No 176 in 1957. The latter — and other BUT railcar conversions — have the rubber bellows type of gangway which distinguishes the 1957 BUT railcars. No 9, a later 60ft vehicle, was converted to run with the 1950/1 diesel sets and has a driving cab. No 175 is a 58ft BUT railcar trailer conversion seating 20 and has a train heating boiler, as has No 231. Nos 227/231 are 60ft 'Enterprise' vehicles. The former seats 50, the latter 18. Both were converted for railcar use in 1957. The 1959 numbers are those in brackets.

Nos: 176 (583), 98 (581). **Type:** Third Open (1941). **Owner:** Railway Preservation Soc of Ireland. **Location:** Whitehead.

No: 175 (594). **Type:** Third Brake Corridor (1946). **Owner:** Railway Preservation Soc of Ireland. **Location:** Whitehead.

No: 227 (561). **Type:** First Corridor (1947). **Owner:** Railway Preservation Soc of Ireland. **Location:** Whitehead.

No: 231 (562). **Type:** First Brake Open (1948). **Owner:** Railway Preservation Soc of Ireland. **Location:** Whitehead.

No: 9 (586). **Type:** Third Open (1954). **Owner:** Railway Preserevation Soc of Ireland. **Location:** Whitehead.

London Midland & Scottish Railway
Engineers' Saloon   1944

This coach is one of 14 inspection saloons built 1940-7. The steel-panelled body is 50ft long and 9ft 3in wide with observation windows at each end. The interior layout consists of two saloons interspersed by a central lavatory/kitchen/guard's section. It came to the Severn Valley in 1972 via the Standard Gauge Steam Trust at Tyseley and will shortly move again to the Strathspey Railway.

No: 45021. **Type:** District Engineer's Saloon. **Owner:** Private. **Location:** Bewdley.

London Midland & Scottish Railway
Standard Corridor Stock   1945

The post-war LMS corridor stock followed the pre-war type *(page 154)* very closely. The main differences were that the roofs were flush-welded, there were extra doors on the corridor side and those built after 1947 had a 'porthole' window to the lavatory and on the corresponding corridor side. The body dimensions of 57ft x 9ft 3in, 9ft wheelbase bogies and wood/steel construction were maintained. All were withdrawn by 1969.

Of those preserved, the thirds have seven compartments, the brake thirds have four compartments and the open thirds seat 56 passengers in eight bays. A large proportion of these coaches have been restored or are in service. Brake third No 26680 was purchased by the SVR for spares only. Nos 12992/13045 were built by Birmingham RCW, the rest at Derby or Wolverton.

Side-Corridor Stock

No: 12992. **Type:** Third (1949). **Owner:** LMSR & BR Coach Fund. **Location:** Bewdley.

No: 13045. **Type:** Third (1950). **Owner:** Private. **Location:** Bewdley.

No: 26668. **Type:** Third Brake (1950). **Owner:** Private. **Location:** Bewdley.

No: 26921. **Type:** Third Brake (1950). **Owner:** Severn Valley Rly Co Ltd. **Location:** Bewdley.

No: 26986. **Type:** Third Brake (1950). **Owner:** Private. **Location:** Bewdley.

No: 27023. **Type:** Third Brake (1950). **Owner:** LMSR & BR Coach Fund. **Location:** Bewdley.

No: 27043. **Type:** Third Brake (1950). **Owner:** Strathspey Rly. **Location:** Boat of Garten.

Open Stock

No: 27218. **Type:** Third Open. (1945/6). **Owner:** Warwickshire Rly Soc. **Location:** Bewdley.

No: 27220. **Type:** Third Open (1945/6). **Owner:** Severn Valley Rly Co Ltd. **Location:** Bewdley.

No: 27234. **Type:** Third Open (1945/6). **Owner:** Private. **Location:** Boat of Garten.

No: 27249. **Type:** Third Open (1946). **Owner:** Foxfield Lt Rly Soc Ltd. **Location:** Blyth Bridge.

No: 27270. **Type:** Third Open (1947). **Owner:** Private. **Location:** Bewdley.

Nos: 27389, 27407. **Type:** Third Open (1947). **Owner:** Scottish RPS. **Location:** Falkirk.

Southern Railway
Standard Corridor Stock   1947

This Southern corridor stock was probably the best of the post-war designs produced by the 'Big Four.' The first SR post-war coaches appeared in 1945, but these were largely an updated version of

No 5761 as purchased for preservation. R.J. Wigley.

pre-war stock. In late 1945 the standard type began to enter service. These had 63½ ft underframes, 64 ft 6 in over the body ends, and ran on the standard SR 8 ft bogies. The bodysides of 9 ft 3 in width had an attractive curved profile, later adopted for the BR standard stock, but the bodies were of conventional wood-framed, steel-panelled construction with wooden roofs. Compartment access was from the corridor only, with adequate transverse corridors. The bodyside windows had radiused corners and sliding ventilators. These coaches were built until 1951, later examples having deeper sliding ventilators. The brake thirds, such as Nos 2515/4279, had open saloon and compartment seating. The majority of this stock was formed in three or four-coach sets. All disappeared by 1969, following the Bournemouth electrification, some like Nos 1469/82 having migrated to the Eastern or Scottish Regions.

Nos 5761/8 are early composites with four first and three third-class compartments. Nos 4279/2515 have two compartments and a centre gangway saloon with four bays; No 4279 was built by Birmingham RCW and has some external differences. The open thirds have eight bays, seating 64.

Side-Corridor

No: 5761. **Type:** Composite (1947). **Owner:** South Eastern Steam Centre. **Location:** Ashford (Kent).

No: 5768.**Type:** Composite (1947). **Owner:** The Bulleid Soc Ltd. **Location:** Sheffield Park.

No: 4279. **Type:** Third Brake (1948). **Owner:** Bluebell RPS. **Location:** Sheffield Park.

No: 2515. **Type:** Third Brake (1951). **Owner:** Southern Loco Pres Co Ltd. **Location:** Sheffield Park.

Open

No: 1464. **Type:** Third (1951). **Owner:** Merchant Navy Loco Pres Soc. **Location:** Ashford (Kent).

No: 1469. **Type:** Third (1951). **Owner:** Vintage Carriages Trust. **Location:** Haworth.

No: 1481. **Type:** Third (1951). **Owner:** Bluebell RPS. **Location:** Sheffield Park.

No: 1482. **Type:** Third (1951). **Owner:** Southern Loco Pres Co Ltd. **Location:** Sheffield Park.

London & North Eastern Railway
Non-corridor Stock   1947

The LNER design post-war non-corridor stock was built on a 62 ft 2½ in underframe and had steel body panelling on wooden framing. The first examples appeared in 1947 and construction continued until 1953. Those built after 1949 had rounded corners to the windows. Withdrawals began in 1962-3 and nearly all were taken out of service by 1967.

No 88329 is from one of the 1947 batches and was built by Cravens Ltd. It has three first and four third-class compartments with centre lavatories and an internal side corridor to the first-class compartments. It was withdrawn from the Scottish Region in 1967 and was overhauled at York before coming on to the North Yorkshire Moors Railway. It is

No 88329 (first coach) in use on the North York Moors Rly. P.J. Robinson.

now only used occasionally on passenger trains.

No 80417 is a brake composite with two first-class and four third-class compartments built by R. Y. Pickering & Co Ltd in 1952. It spent all its life in the Scottish Region and was acquired in 1969. It has been used on SRPS railtours over BR metals in the past but, being non-corridor, it is unlikely to be used on long tours again.

No: 88329. **Type:** Lavatory Composite (1947). **Owner:** North York Moors Rly. **Location:** Grosmont.
No: 80417. **Type:** Brake Composite (1952). **Owner:** Scottish RPS. **Location:** Falkirk.

Pullman Car Co
'Devon Belle' Observation Saloon   1947

The 'Devon Belle' was the only new prestige train of the early post-war period. It ran from June 1947 until the end of the 1955 summer timetable, operating during the summer only between Waterloo and Ilfracombe and, until 1950, to Plymouth. A set of fairly old, reconditioned Pullman cars was used but the two observation cars, Nos 13/4, had new 58 ft 6 in by 8 ft 7 in bodies on underframes which had originally been under LNWR ambulance coaches, later converted to Pullman cars. Really excellent visibility was (and is) given through the large observation windows, which were double-glazed. Twenty-seven passengers were accommodated in armchairs and double settees which could be swivelled to any angle on fixed pivots. At the train

No 13 in use as M 280, 1960. J.B. Bucknell.

end of the car was a kitchen/bar. The interior was attractively panelled in plastics with a large map of Devon on the kitchen bulkhead. After 1955 the observation cars had no regular employment until sold in 1958. No 13 was sold to the LMR, renumbered M280 and put to work on the North Wales land cruise train. The other car as SC281 went to the Scottish Region. Their new livery was in similar style to the Pullman scheme but with maroon paint in place of Pullman brown. They were withdrawn in the mid-1960s. No 281 went over to the USA with *Flying Scotsman* and has not returned.

No: 13. **Type:** Observation Car. **Owner:** Torbay Steam Rly. **Location:** Paignton.

Great Western Railway
Standard Corridor Stock 1948

The GWR set about designing new corridor stock in 1944, contemplating such innovations as aluminium construction, plastic interior panelling and fluorescent lighting. In the event, although some of these features appeared in one or two coaches, shortage of materials precluded their general use — and delivery, in any case, was much delayed. Quite a number were built by outside contractors, such as Nos 829/2119 (Gloucester RCW) and 2202 (Metro-Cammell).

The coaches had wooden body framing and steel panelling. The bodies were 63 ft or 64 ft in length and 8 ft 11 in wide. The outward appearance established a new outline for the GWR as the bodysides were slab-sided and the roofs sloped down at each end. All ran on 9 ft wheelbase bogies. Coaches of this design appeared between 1946 and 1951 and were all withdrawn by the end of 1967.

The brake composites have two first and four third-class compartments. They were allocated for Royal Train use and probably never

Third No 855 in GWR livery.

184

ran in general traffic. Nos 829/2119 are eight-compartment thirds and No 2202 is a four-compartment brake third. All five coaches are serviceable and those on the SVR, although in service, have not yet been fully restored to original livery.

No: 7372. **Type:** Brake Composite (1948). **Owner:** Great Western Soc Ltd. **Location:** Bodmin.
No: 7377. **Type:** Brake Composite (1948). **Owner:** Dart Valley Lt Rly. **Location:** Buckfastleigh.
No: 829. **Type:** Third (1945). **Owner:** Great Western (SVR) Assoc. **Location:** Bewdley.
No: 2119. **Type:** Third (1949). **Owner:** Private. **Location:** Bewdley.
No: 2202. **Type:** Third Brake (1950). **Owner:** Private. **Location:** Didcot.

Great Western Railway
Inspection Saloons   1948

No 80972 as restored. D.A. Idle.

As described on page 103, nearly all the GWR engineers' inspection saloons were converted from old vehicles. In 1948 some very ancient four-wheelers still survived and six new 52 ft purpose-built saloons were constructed to replace them. These had the general profile of the 1936-41 corridor stock and had observation windows at each end. The interior consisted of a saloon at each end and the usual central lavatory and pantry. Standard 9ft wheelbase bogies were used. Warning gongs were fitted at each end and folding steps were provided at each side. Probably all were repainted in chocolate and cream livery after 1956 and two of the three preserved were withdrawn in this condition. Nos 80969/72/4 were taken out of service in 1972/3.

No 80969 was originally allocated to the Newport Division, but was latterly in Scottish Region stock. No 80972 was sent to Shrewsbury when new and 80974 was at Wolverhampton.

No: 80969. **Type:** Inspection Saloon. **Owner:** Private. **Location:** Bewdley.
No: 80972. **Type:** Inspection Saloon. **Owner:** Birmingham Rly Museum. **Location:** Tyseley.
No: 80974. **Type:** Inspection Saloon. **Owner:** Private. **Location:** Grosmont.

British Railways (SR)
4-DD Electric Stock   1949

These three coaches are the survivors from one of the two 1949-built SR 'double-decker' electric multiple-units — originally Nos 4001/2, later 4901/2. The 'double-decker' 4-DD four-coach trains were an atempt to alleviate overcrowding on the Eastern Division of the SR, but a better solution was found in lengthening platforms and running longer trains. Each 4-DD could seat 508 passengers, compared to 386 in an ordinary four-car set of similar length. Strictly speaking they were not double-decked but with one high, one low compartment alternatively. The increased overall height — some 4½in greater then normal — meant that their operations were restricted. Their usual haunt was the North Kent line. They could not work with other stock. In October 1971 both sets were withdrawn rather than receiving a general overhaul. Two of the three remaining cars were up for sale in 1975.

Set No 4002 in BR service 1969. J.H. Bird.

No: 13003. **Type:** Motor Second Brake. **Owner:** South Eastern Steam Centre.
Location: Ashford (Kent).
No: 13004. **Type:** Motor Second Brake. **Owner:** South Eastern Steam Centre.
Location: Ashford (Kent).
No: 13503. **Type:** Trailer Second. **Owner:** South Eastern Steam Centre.
Location: Ashford (Kent).

British Railways (LMS)
All-Steel Corridor Stock   1950

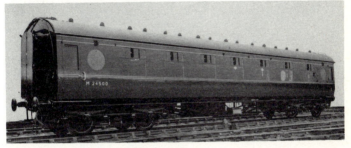

No 24500 as built and in LMS-style livery.

These coaches are of interest as the first all-steel post-war vehicles,
the experience in building them probably contributing to the decision
by BR to standardise on all-steel coaches after 1951. The LMS, in
any case, had built all-steel electric multiple-unit stock from 1938.
The body profile is different to other LMS post-war stock, having
noticeably 'bulging' bottom panels to the body. These coaches also
have 'porthole' windows. The length over the headstocks is 60ft 0¾in
(LMS corridor composites were 60ft as opposed to 57ft) and the body
width 9ft 3in. The interiors have a number of features perpetuated
in BR standard stock.

Two hundred and forty of these composites were built 1949-50
and all were withdrawn by 1969. Both Nos 24617/24725 were
purchased intact and are in use, the latter on specials over BR lines.

No: 24617. **Type:** Composite. **Owner:** Private. **Location:** Bewdley.
No: 24725. **Type:** Composite. **Owner:** Scottish RPS. **Location:** Falkirk.

British Railways (LNER)
Standard Corridor Stock

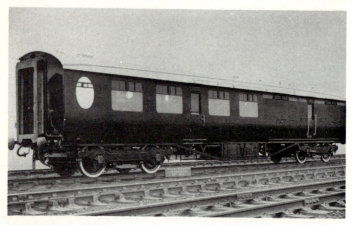

Third Brake No. 1909.

The post-war LNER corridor stock represented a complete break from the teak-panelled Gresley era coaches. The first appeared in 1945 and over 800 were built until 1950. The composites had bodies only 59ft 6in over the ends; all other types were 63ft. The bodies were steel panelled on wooden frames. For each type of corridor stock the interior was divided into two or three compartments separated by transverse vestibules for easy access. The standard 8ft 6in bogies were used for most of this stock. All were withdrawn by 1968.

No 18477 built at York has three first-class and three third-class compartments. It was acquired in 1969 and, although used on NYMR services in early days, is now awaiting complete restoration.

The third brake has four compartments. Its original number not yet known; on withdrawal it was converted by BR as a civil engineers' mess/tool van. Its new owners propose to use it as a dormitory/mess/tool van for use with A2 Pacific No 532 'Blue Peter' and it is passed for main line running.

No: (DE 321120). **Type:** Third Brake. **Owner:** Blue Peter Locomotive Soc. **Location:** Walton Colliery, Wakefield.
No: 18477. **Type:** Composite. **Owner:** North York Moors Rly. **Location:** Grosmont.

British Railways (LNER)
Sleeping Cars   1950

The 65ft underframe was used for the post-war LNER-design sleeping cars which were built at Doncaster. The bodies, 66ft 6in over the ends, were steel panelled. The first-class sleepers had ten compartments and an attendant's compartment, and the third-class cars — the batch Nos 1765-70 — twelve two-tier berths without an attendant's compartment. Standard compound-bolster 8ft 6in bogies were fitted.

Those preserved are all used for volunteer sleeping accommodation. Nos 1259/60 and 1767 were withdrawn from East Coast service in the late 1960s. Nos 1769/70 had latterly been transferred to the Western Region.

No: 1259. **Type:** First Sleeper (1950). **Owner:** North York Moors Rly. **Location:** Grosmont.
No: 1260. **Type:** First Sleeper (1950). **Owner:** Welshpool & Llanlair Lt Rly Pres Co. **Location:** Llanlair Caereinion.
No: 1767. **Type:** Third sleeper (1952). **Owner:** Cranmore Rly Co Ltd. **Location:** Cranmore.
Nos: 1769, 1770. **Type:** Third sleeper (1952). **Owner:** Keighley & Worth Valley RPS. **Location:** Haworth.

No 1767 en route to Cranmore.

CDRJC
Diesel Railcars (3ft gauge) 1950

The two 1950/1 cars were the last to be put into service. They were based on the railcars built in the 1930s but differed in having a full-fronted cab completely enclosing the engine. As before, the cab and saloon were articulated above the power bogie, which had a Gardner 5LW engine. The body, with two pairs of doors on each side, seated 41. The length over the whole body unit was 41ft 2¾in, the width 7ft 6in. The power bogie with cab unit was built by Walker Bros of Wigan, the coach body by the Great Northern Rly works at Dundalk. No 19 went in traffic in January 1950, No 20 a year later. Both ran until the end of 1959 and were then bought by the Isle of Man Railway, arriving at Douglas (IOM) in 1961. The IOMR coupled them back to back as they were single-ended and they were used to some extent until the old company finished in 1965. Since then their use appears to have been sparodic and now, like all the IOMR stock, they await an uncertain future.

Railcar No 18, built in 1940, of similar Walker Bros/GNR Dundalk parentage was bought after the closure of the railway and stored until 1974. It has now been donated to the North-West of Ireland Railway Society for a working museum project at Londonderry.

No: 19. Type: Diesel Railcar (1950). **Owner:** The Isle of Man Rly. **Location:** Douglas (IOM).
No: 20. Type: Diesel Railcar (1951). **Owner:** The Isle of Man Rly. **Location:** Douglas (IOM).
No: 18. Type: Diesel Railcar (1940). **Owner:** North-West of Ireland Rly Soc. **Location:** Londonderry.

Nos 19/20 in use in the Isle of Man. P. J. Sharpe.

No 221 in service 1952. H. C. Casserley.

A series of new auto-trailers appeared in 1951, Nos 220-34, which were 64ft long and 8ft 11in wide. The bodies had a slab-sided profile with large windows. The interiors had all transverse seating. The cars were generally distributed over the WR system and originally appeared in crimson and cream livery. The standard 9ft bogies were used. In 1954, at the time when the other BR Regions were receiving diesel railcars, a further ten auto-trailers were built, Nos 235-44. These were generally similar to the 1951 batch. Increasing dieselisation and branch-line closures sealed the fate of most by 1964, although some lasted into the next year on push-pull services on former Southern territory.

Those preserved have been repainted in GWR-style chocolate and cream livery, which they never carried, but which can be considered justifiable artist's licence. Those on the Dart Valley normally work in push-pull rakes with auto-fitted 0-6-0PTs. No 231 has also been nicely restored and is used on Didcot open days with car No 190 (*page 145*).

Nos: 225, 228, 232. **Type:** Auto-Trailer (1951). **Owner:** Dart Valley Lt Rly. **Location:** Buckfastleigh.
No: 231. **Type:** Auto-Trailer (1951). **Owner:** Great Western Soc Ltd. **Location:** Didcot.
Nos: 238, 240. **Type:** Auto-Trailer (1954). **Owner:** Dart Valley Lt Rly. **Location:** Buckfastleigh.

British Railways (GWR)
Sleeping Cars

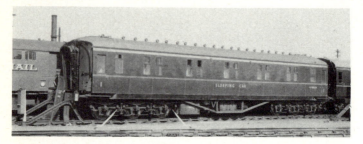

No 9085 in BR Service. R. C. Riley.

Ordered in 1947, the four post-war GWR first-class sleeping cars were not completed until 1951. They were built to the same outline and dimensions as the corridor stock (*page 184*), but had six-wheel 11ft 6in wheelbase bogies. Ten sleeping berths, attendant's compartment and lavatory were accommodated in the 64ft body, which was equipped with pressure ventilation. All four entered traffic in BR crimson and cream livery and were withdrawn in 1970. The whole batch was purchased for preservation, to be used as sleeping accommodation for volunteer workers at Didcot and on the SVR.

Nos: 9082, 9085. **Type:** First Sleeper. **Owner:** Severn Valley Rly Co Ltd. **Location:** Bewdley.
No: 9083. **Type:** First Sleeper. **Owner:** Great Western Soc Ltd. **Location:** Didcot.
No: 9084. **Type:** First Sleeper. **Owner:** Private. **Location:** Bewdley.

British Railways (LMS)
Sleeping Car
1951

A twelve-wheel 69ft body first-class sleeping car design was introduced by the LMS in 1935 and this modern flush-panelled type with pressure ventilation was built in 1951/2. These later cars were among the heaviest non-electric coaches, with a weight of 47 tons. Six-wheel bogies were fitted, these vehicles being the last new twelve-wheeled coaches to be built in Britain. The interiors have twelve berths and an attendant's compartment. All were in service until 1966; many received the blue/grey livery and they survived in traffic until the early 1970s.

No 394 has been restored in LMS livery — which it never carried. No 395 has been repainted in the correct scheme of BR crimson/cream paint and has been passed by BR for mainline running. No 398 is to be used for volunteer sleeping accommodation.

No: 394. **Type:** First Sleeper. **Owner:** Strathspey Rly. **Location:** Boat of Garten.

No: 395. **Type:** First Sleeper. **Owner:** Mr. W. H. McAlpine/Flying Scotsman Enterprises. **Location:** Carnforth.

No: 398. **Type:** First Sleeper. **Owner:** Bluebell RPS. **Location:** Sheffield Park.

Coach No. 398 as purchased. T.J. Edgington.

British Railways
Standard Corridor Stock   1951

A large number of coaches were built to pre-nationalisation designs after 1948, as already described. The BR standard corridor stock first appeared in late 1950 following design work over a period at several works. The new coaches represented a complete break with existing practice, the body being of all-welded steel construction. The underframes with buckeye couplers were 63ft 5in, the slightly bow-ended body being 64ft 6in long and 9ft 3in wide. The bogies were of a new double-bolster design with an 8ft 6in wheelbase. This stock was built in quantity from 1951 until 1964 for locomotive-hauled services. Electric multiple-units for the SR with the same general design were constructed until 1972. A lot of this Mark I stock is still in traffic.

With so many vehicles in private ownership, detailed descriptions cannot be given for reasons of space. A number have been painted in the liveries of their owners and carry new numbers. The prefix indicating Regional allocation is the original one when built; this also applies to BR stock described on pages 196/197.

Open Second (8 bays, 64 seats).

No: S3825. **Built:** 1953. **Owner:** Keighley & Worth Valley Rly. **Location:** Haworth.

No: E3860. **Built:** 1953. **Owner:** North York Moors Rly. **Location:** Grosmont.

No: E3868. **Built** 1953. **Owner:** North Norfolk Rly. **Location:** Sheringham.

No: S3925. **Built:** 1954. **Owner:** Conwy Valley Rly Museum. **Location:** Bettws-y-Coed.

Nos: S4046, E4166, E4275, E4288, E4289, E4317. **Built:** 1956. **Owner:** Dart Valley/Torbay Steam Rly. **Location:** Buckfastleigh/Paignton.

No: E4127. **Built:** 1956. **Owner:** Strathspey Rly. **Location:** Boat of Garten.

No: E4280. **Built:** 1956. **Owner:** Keighley & Worth Valley Rly. **Location:** Haworth.

No: E4286. **Built:** 1956. **Owner:** National Rly Museum, York.

No: E4350. **Built:** 1956. **Owner:** Bury Transport Museum. **Location:** Bury.

No: M4481. **Built:** 1957. **Owner:** North York Moors Rly. **Location:** Grosmont.

No: E4496. **Built:** 1956/7. **Owner:** David & Charles Holdings Ltd. **Location:** Newton Abbot.

Nos: E4507, E4564. **Built:** 1956/7. **Owner:** Dart Valley/Torbay Steam Rly. **Location:** Buckfastleigh/Paignton.

No: E4597. **Built:** 1956/7. **Owner:** North York Moors Rly. **Location:** Grosmont.

No: E4654. **Built:** 1957. **Owner:** Dart Valley/Torbay Steam Rly. **Location:** Buckfastleigh/Paignton.

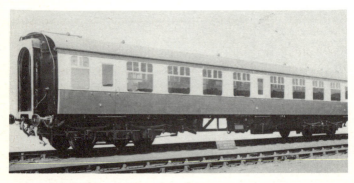

Open second E4099 as built.

Brake Open Second (5 bays, 39 seats).
No: E9218. Built: 1955/6. **Owner:** Lakeside Rly Soc. **Location:** Haverthwaite.
No: E9220. Built: 1955/6. **Owner:** Private. **Location:** Bridgnorth.
Nos: E9235, W9267. Built: 1955/6. **Owner:** North York Moors Rly. **Location:** Grosmont.
No: E9241. Built: 1955/6. **Owner:** Cranmore Rly. **Location:** Cranmore.
No: E9315. Built: 1956. **Owner:** Lochty Private Rly. **Location:** Lochty.
No: SC9362. Built: 1956. **Owner:** Strathspey Rly. **Location:** Boat of Garten.

Second (8 compartments, 64 seats).
No: E24248. Built: 1951/2. **Owner:** Lakeside Rly Soc. **Location:** Haverthwaite.
Nos: W24377, W24381, E24449, SC24731. Built: 1953. **Owner:** Lakeside Rly Soc. **Location:** Haverthwaite.
Nos: SC24396, SC24726. Built: 1953. **Owner:** LMSR BR Coach Fund. **Location:** Bridgnorth.
No: M24412. Built: 1953. **Owner:** Severn Valley Rly. **Location:** Bridgnorth.
No: E24799. Built: 1954. **Owner:** Lakeside Rly Soc. **Location:** Haverthwaite.

First (7 compartments, 42 seats).
No: E13043. Built: 1951. **Owner:** North York Moors Rly. **Location:** Grosmont.
No: E13102. Built: 1954. **Owner:** North York Moors Rly. **Location:** Grosmont.

Brake Second (4 compartments, 32 seats).
No: E34181. Built: 1952. **Owner:** Lakeside Rly Soc. **Location:** Haverthwaite.
No: E34427. Built: 1953. **Owner:** Severn Valley Rly. **Location:** Bridgnorth.
No: M34460. Built: 1953. **Owner:** Dinting Rly Centre. **Location:** Dinting (as mobile shop).
No: M34562. Built: 1955, **Owner:** Private. **Location:** Bridgnorth.
No: E34612. Built: 1955. **Owner:** Mr W.H. McAlpine/Flying Scotsman Enterprises. **Location:** Carnforth (as administration vehicle).

Composite (Four 1st, three 2nd, 2 x 24 seats).
No: W15096. Built: 1951. **Owner:** Main Line Steam Trust. **Location:** Loughborough.
No: E15401. Built: 1954. **Owner:** Strathspey Rly. **Location:** Boat of Garten.
No: E15514. Built: 1954. **Owner:** Peterborough Rly Socy. **Location:** Wansford.
No: SC15553. Built: 1954. **Owner:** Private. **Location:** Bridgnorth.
Nos: S15566, S15577. Built: 1954. **Owner:** Great Western Soc. **Location:** Didcot (as sleeping accommodation).
No: W15611. Built: 1955. **Owner:** Main Line Steam Trust. **Location:** Loughborough.
No: M15829. Built: 1956. **Owner:** 6000 Locomotive Assoc. **Location:** Hereford.

Pullman Car Co
Pullman Cars

Carina *as built for 'Golden Arrow' service.*

The 'Golden Arrow' had only a brief period of glory in post-war days and its all-Pullman formation gradually gave way to the inclusion of ordinary coaches. Ten new Pullman cars were built in 1951-2 using materials which had been ordered before the war. Seven were built by Birmingham RCW, including *Aquila* and *Orion*, and the others by Pullman at Brighton. The cars were a new design, 63ft 10in long and 8ft 5½in wide, generally similar to the 1928 cars but distinguishable by a squarish lavatory window and LNER-pattern double-bolster bogies. The interiors were a slightly modernised version of the traditional Pullman specification. The 1951-2 cars were used on the 'Golden Arrow' and 'Bournemouth Belle,' working on the former until October 1972. Some had already been withdrawn, such as *Aquila*. The last five survivors, including *Orion*, had been repainted in BR blue/grey livery. *Carina* and *Phoenix* have been sold for export.

Name: *Aquila*. **Type:** First Kitchen. **Owner:** H.P. Bulmer Ltd. **Location:** Hereford.

Name: *Orion*. **Type:** First Kitchen. **Owner:** Private. **Location:** Ashford (Kent).

Name: *Perseus (301)*. **Type:** First Parlour. **Owner:** Scottish & Newcastle Breweries Ltd.

Name: *Cygnus (308)*. **Type:** First Parlour. **Owner:** Scottish & Newcastle Breweries Ltd.

British Railways
Standard Non-corridor Stock

The BR standard non-corridor coaches entered service from 1954. As with the corridor stock these vehicles had bodies of welded all-steel construction. Most were built on a short 56ft 11in underframe; none of those with the 63ft underframe has been preserved. The BR 8ft 6in wheelbase bogie was used. The standard non-corridor stock was built from 1954-6 and the majority of coaches had a very short life, being withdrawn from 1962, although 94 coaches survived in late 1973 on Great Northern suburban services; these were to be withdrawn in 1976/7. All this stock was screw-coupled.

The seconds and brake seconds are of usual compartment layout. The lavatory composites have internal side corridors to the compartments, with lavatories dividing first and third-class. There are long windows to the corridor sides. The open seconds have a centre gangway layout with lavatories in the middle of the body. Brake Third No 43181, formerly on the Keighley and Worth Valley Rly, has now been scrapped.

No 1000 is a unique vehicle. The body is constructed from glass-reinforced plastics on a 63ft 5in underframe from a BR standard corridor coach destroyed in the Lewisham crash of 1957. It has ten compartments, each with different decor scheme. It was built in 1962 and appeared in July 1963 as DS 70200 and was used in the Lancing works train. After a spell on the Hayling Island branch (by now numbered S1000) it was transferred to the Clapham Junction-Kensington service in 1964. It was withdrawn in 1967 and remained out of use for a long while.

Lavatory Composite (Three first, Five second).
No: E43010. **Built:** 1954/5. **Owner:** Bluebell Rly. **Location:** Sheffield Park.
No: E43024. **Built:** 1954/5. **Owner:** Strathspey Rly. **Location:** Boat of Garten.
Nos: E43034, E43041. **Built:** 1954/5. **Owner:** North Norfolk Rly. **Location:** Sheringham.

Composite No 43005 in BR service. G. M. Kichenside.

SO (8 bays, 80 seats).
No: E48004. **Built:** 1955. **Owner:** Derby Corporation. **Location:** Butterley.
No: E48007. **Built:** 1955. **Owner:** Somerset & Dorset Museum Trust. **Location:** Taunton.
Nos: E48010, E48011. **Built:** 1955. **Owner:** Keighley & Worth Valley Rly. **Location:** Haworth.
No: E48015. **Built:** 1955. **Owner:** Bluebell RPS. **Location:** Sheffield Park.
No: E48026. **Built:** 1955. **Owner:** North Norfolk Rly. **Location:** Sheringham.

BS (6 compartments).
No: E43147. **Built:** 1954/5. **Owner:** Stour Valley RPS. **Location:** Chappel.
No: E43172. **Built:** 1954. **Owner:** Bluebell Rly. **Location:** Sheffield Park.
No: E43186. **Built:** 1954. **Owner:** Derby Corporation. **Location:** Butterley.
Nos: M43231, M43289. **Built:** 1954. **Owner:** Cranmore Rly. **Location:** Cranmore.
No: E43345. **Built:** 1955. **Owner:** Keighley & Worth Valley Rly. **Location:** Haworth.
No: E43349. **Built:** 1955. **Owner:** Strathspey Rly. **Location:** Boat of Garten.

S (9 compartments).
No: E46093. **Built:** 1954. **Owner:** Somerset & Dorset Museum Trust. **Location:** Taunton.
No: E46097. **Built:** 1954. **Owner:** Derby Corporation. **Location:** Butterley.
No: W46139. **Built:** 1954. **Owner:** Stour Valley RPS. **Location:** Chappel.
No: W46142. **Built:** 1954. **Owner:** North York Moors Rly. **Location:** Grosmont.
Nos: W46145, W46157. **Built:** 1954. **Owner:** Keighley & Worth Valley Rly. **Location:** Haworth.
No: E46228. **Built:** 1954/5. **Owner:** Keighley & Worth Valley Rly. **Location:** Haworth.
No: E46235. **Built:** 1954/5. **Owner:** North York 'Moors Rly. **Location:** Grosmont.
No: S1000. **Built:** 1962. **Owner:** Cranmore Rly. **Location:** Cranmore.

British Railways
Experimental Stock   1957

As part of the later phase of the Modernisation Plan, BR commissioned private contractors and Doncaster works to produce 14 prototype first-and second-class main-line coaches during 1957. The interior layout and decor of these vehicles was left entirely to the builders, who had to keep within the limits of the standard BR corridor stock bodyshell. It was intended that public reaction to the features of the prototypes would determine the development of a new range of BR stock, but in the event a completely fresh start was

No 3083 as built.

made by BR in 1964 with the XP64 prototype train. The majority were allocated to the Eastern Region but operated on other Regions in time. By late 1973 only four of the prototypes remained in service.

No 3081 was built by Birmingham RCW with interior design consultancy from Sir Hugh Casson. It has 33 individual reclining seats with much wider bays than normal, rather poor visibility from the 5½ft width bodyside windows, double glazing and indirect fluorescent lighting. No 3083/4 were Doncaster's contribution to open first designs. No 3083 is the more interesting of the two, with 36 revolving and reclining individual seats and 12 narrow windows to each bodyside. No 3084 seats 36 but in individual armchair seats. These coaches finished their BR days on WR commuter trains until withdrawn in 1972. No 3785 was built in 1953 but refurbished in 1957 to become one of the 'new look' prototypes.

Nos: E3081, E3084. **Type:** Open First. **Owner:** Dart Valley Lt Rly. **Location:** Paignton.

No: E3083. **Type:** Open First. **Owner:** Severn Valley Rly Co Ltd. **Location:** Bewdley.

No: E3785. **Type:** Open Third. **Owner:** Severn Valley Rly Co Ltd. **Location:** Bridgnorth.

British Railways
Diesel Multiple-units   1957-61

The Glasgow-Edinburgh service was worked by six-car diesel multiple-units from 1956-70. Apart from four cars all this stock has now been scrapped or sold out of service. These buffet firsts — body 64ft 6in by 9ft 3in — have three first-class compartments with side-corridor,

6-car Glasgow-Edinburgh set. J.M. Boyes.

a buffet and a kitchen. No 79441 has been painted in Strathspey Rly livery and named *"Glenfiddich."*

The two-car Gloucester RCW-built railcars and one trailer are from an order delivered from 1957 for Edinburgh suburban services; some later went to the Eastern Region. They are of lightweight steel tubular-framed integral construction. The bodies are 57ft 6in by 9ft 3in. Motor cars seat 52, trailers twelve first-class, eighteen second-class. The North York Moors sets are painted in LNER-style cream and green livery and work the Grosmont-Pickering trains.

The Park Royal cars are also of lightweight integral construction. The bodies are also 57ft 6in by 9ft 3in. The motor cars seat 52 and the trailers originally 16 first-class, 48 third-class. They are intended for use on the Taunton-Minehead line, yet to be reopened.

One Metro-Cammell buffet car trailer has been bought by the KWVR out of a batch of six built for North Eastern Region semi-fast services. The body is 57ft by 9ft 3in and it seats 53; the small buffet is at one end.

Glasgow-Edinburgh Inter-City 2 units (built 1956-61) Swindon/AEC.

No: SC59098. **Type:** First Trailer Buffet (1961). **Owner:** Strathspey Rly.
   Location: Boat of Garten.
No: SC59099. **Type:** First Trailer Buffet (1961). **Owner:** Strathspey Rly.
   Location: Boat of Garten.
No: SC79441. **Type:** First Trailer Buffet (1956). **Owner:** Strathspey Rly.
   Location: Boat of Garten.
No: SC79443. **Type:** First Trailer Buffet (1956). **Owner:** North York Moors Rly.
   Location: Grosmont.

Gloucester RCW/AEC Engine 2-car units (built 1957).

No: SC50341. **Type:** Second Motor Brake. **Owner:** North York Moors Rly.
   Location: Grosmont.

No: SC56099. **Type:** Driving Trailer Composite. **Owner:** North York Moors Rly. **Location:** Grosmont.

No: SC51118. **Type:** Second Motor Brake. **Owner:** North York Moors Rly. **Location:** Grosmont.

No: SC56097. **Type:** Driving Trailer Composite. **Owner:** North York Moors Rly. **Location:** Grosmont.

No: SC56301. **Type:** Driving Trailer Composite. **Owner:** RPS Chasewater Lt Rly. **Location:** Brownhills.

Park Royal/AEC 2-car units (built 1958).
No: M50413. **Type:** Second Motor Brake. **Owner:** West Somerset Rly Co Ltd. **Location:** Taunton.

No: M56168. **Type:** Driving Trailer Composite. **Owner:** West Somerset Rly Co Ltd. **Location:** Taunton.

No: M50414. **Type:** Second Motor Brake. **Owner:** West Somerset Rly Co Ltd. **Location:** Taunton.

No: M56169. **Type:** Driving Trailer Composite. **Owner:** West Somerset Rly Co Ltd. **Location:** Taunton.

Metro-Cammell Ltd Trailer (built 1960).
No: E59575. **Type:** Second Trailer Buffet. **Owner:** Keighley & Worth Valley RPS. **Location:** Haworth.

British Railways
Four-Wheel Diesel Railbus   1958

BR introduced four-wheel diesel railbuses for lightly trafficked branch lines but before long these services were closed under the

No 79970 in service 1966 — type not preserved. D. Cross.

Beeching Plan and few were in service after 1964/5. The German-built cars — 41ft 10in over the body — went into service from Cambridge depot but were always plagued by spares shortages. The body has a centre entrance vestibule with power-operated doors. The interior seats 56 in two saloons. The AC Cars' railbuses are 37ft long, seat 46 and also have air-operated doors. These cars were used on WR Kemble-Tetbury/Cirencester services. The KWVR cars are used on 'off-peak' services.

Nos: E79960, E79962. **Type:** Waggon und Maschinenbau/Buessing. **Owner:** North Norfolk Rly. **Location:** Sheringham.
Nos: E79963, E79964. **Type:** Waggon und Maschinenbau/Buessing. **Owner:** Keighley & Worth Valley Rly. **Location:** Haworth.
No: W79976. **Type:** AC Cars/AEC. **Owner:** Somerset Rly Museum. **Location:** Bleadon & Uphill.
No: W79978. **Type:** AC Cars/AEC. **Owner:** North York Moors Rly. **Location:** Grosmont.

Other Passenger Stock (Non-passenger carrying)

GWR
No: 1399. **Type:** Milk train brake van. **Wheels:** 4-whl. **Built:** 1921. **Location:** Bewdley.
No: 2796. **Type:** Siphon 'G'. **Wheels:** Bogie. **Built:** 1937. **Location:** Didcot.
No: 752. **Type:** Special cattle van. **Wheels:** 6-whl. **Location:** Didcot.
No: 765. **Type:** Special cattle van. **Wheels:** 6-whl. **Built:** 1953. **Location:** Chappel (Essex).

SR (pre-Grouping)
No: 1450. **Type:** SECR carriage van. **Wheels:** 6-whl. **Location:** Ashford (Kent).
No: (1541). **Type:** LSW milk van. **Wheels:** 6-whl. **Location:** Quainton Road.
No: (DS1686). **Type:** LSW milk van. **Wheels:** 4-whl. **Location:** Sheffield Park.
No: 270. **Type:** LBSC milk van. **Wheels:** 6-whl. **Built:** 1908. **Location:** Sheffield Park.
No: 1601. **Type:** SECR pass bk van. **Wheels:** 6-whl. **Built:** 1910. **Location:** Sheffield Park.

SR (post-Grouping)
No: 404. **Type:** Full brake (BY). **Wheels:** 4-whl. **Location:** Ashford (Kent).
No: 407. **Type:** Full brake (BY). **Wheels:** 4-whl. **Built:** 1937. **Location:**
No: 407. **Type:** Full brake (BY). **Wheels:** 4-whl. **Location:** Bettws-y-Coed.
No: 442. **Type:** Full brake (BY). **Wheels:** 4-whl. **Location:** Sheffield Park.
No: 653. **Type:** Full brake (BY). **Wheels:** 4-whl. **Location:** Sheffield Park.
The above built 1937-41.

No: 1070. **Type:** Parcels & Misc. Van (PMV). **Wheels:** 4-whl. **Location:** Bettws-y-Coed.

No: 1108. **Type:** Parcels & Misc. Van (PMV). **Wheels:** 4-whl. **Location:** Quainton Road.

No: 1125. **Type:** Parcels & Misc. Van (PMV). **Wheels:** 4-whl. **Location:** Haworth.

No: 1137. **Type:** Parcels & Misc. Van (PMV). **Wheels:** 4-whl. **Location:** Cranmore.

No: 1396. **Type:** Parcels & Misc. Van (PMV). **Wheels:** 4-whl. **Location:** Cranmore.

No: (DW70031). **Type:** Parcels & Misc. Van (PMV). **Wheels:** 4-whl. **Location:** Sheffield Park. The above built 1936-41.

No: 1990. **Type:** Covered Carriage Truck (CCT). **Wheels:** Bogie. **Built:** 1951. **Location:** Wansford.

No: 2298. **Type:** Covered Carriage Truck (CCT). **Wheels:** bogie. **Built:** 1951. **Location:** Sheffield Park.

LMS (pre-Grouping)

No: 5. **Type:** HR passenger brake van. **Wheels:** 4-whl. **Built:** c1870. **Location:** Boat of Garten.

No: (DM284290). **Type:** GSWR full brake. **Wheels:** 6-whl. **Built:** 1901. **Location:** Falkirk.

No: (DM395106). **Type:** MR motor car van. **Wheels:** 4-whl. **Built:** 1916. **Location:** Butterley.

Type: LNWR parcels van. **Wheels:** 6-whl. **Location:** Wansford.

No: (DM395017). **Type:** LNWR ex-sleeper. **Wheels:** 12-whl. Bogie. **Built:** 1907. **Location:** Quainton Road. Present condition is as cinema car.

No: (DM395273). **Type:** LNWR CCT. **Wheels:** 6-whl. **Built:** 1921. **Location:** Bewdley.

LMS (post-Grouping)

No: 37518. **Type:** CCT (Scenery Van). **Wheels:** Bogie. **Built:** 1927. **Location:** Blythe Bridge.

No: 37519. **Type:** CCT (Scenery Van). **Wheels:** Bogie. **Built:** 1927. **Location:** Blythe Bridge.

No: 38268. **Type:** CCT (Scenery Van). **Wheels:** Bogie. **Location:** Blythe Bridge.

No: 32918. **Type:** Full brake (BZ). **Wheels:** 6-whl. **Built:** 1932. **Location:** Ashchurch.

No: 32919. **Type:** Full brake (BZ). **Wheels:** 6-whl. **Built:** 1932. **Location:** Bewdley.

No: 32988. **Type:** Full brake (BZ). **Wheels:** 6-whl. **Built:** 1938. **Location:** Boat of Garten.

No: 33003. **Type:** Full brake (BZ). **Wheels:** 6-whl. **Built:** 1940. **Location:** Boat of Garten.

No: 33016. **Type:** Full brake (BZ). **Wheels:** 6-whl. **Built:** 1940. **Location:** Bettws-y-Coed.

No: 40226. **Type:** Fish van. **Wheels:** 6-whl. **Built:** 1946. **Location:** Falkirk.

Type: Full brake (BY). **Wheels:** 4-whl. **Built:** 1929. **Location:** Sheringham.
No: 1370. **Type:** CCT. **Wheels:** Bogie. **Built:** 1950. **Location:** Bettws-y-Coed.

BR

No: 96369. **Type:** Horse box. **Wheels:** 4-whl. **Built:** 1957. Department of Education & Science (in store).

No: 96403. **Type:** Horse box. **Wheels:** 4-whl. **Built:** 1957. **Location:** Quainton Road.

In addition there are a few coach bodies preserved which deserve mention. What is thought to be a North Staffordshire Railway 25ft luggage second built c 1870-90 has been acquired by the North Staffordshire Railway Society. An ex-London Tilbury & Southend Rly 25ft third built 1876 has been purchased by a private owner at Pitsea. The Steam Centre at Kirkmichael, Isle of Man, has two Isle of Man Railway four-wheel coach bodies dating from 1873. London Transport have recently discovered the body of a first-class 'Jubilee' stock coach of 1887 in a farmyard in Berkshire. As yet no decision has been taken on whether it will be preserved in whole or in part. Another coach body of interest is that of a third-class coach of about 1840 from one of the North Eastern's constituents which was at the old York Railway Museum.

As mentioned in the Introduction, no compete GWR broad gauge coach has been preserved. However, the Bristol Museum has parts of two or three coach bodies and hopes at some time in the future to reconstruct one complete body. One of the Bristol Museum's acquisitions is the side of an ex-Bristol & Exeter four-compartment first of c 1849 which bears evidence of the GWR livery of the 1880s.

There is also a Wagon-lit coach, No. 2757, a third-class sleeping car of 1926 which has been incorporated into a restaurant at the 'Denham Express' public house at Denham, Bucks.

List of Preservation Society Sites and Museums*

In most cases the headquarters of the organisation is given and the vehicles may actually be located elsewhere or be in transit. Some coaches may be in process of restoration and not be on view to the public.

Ashchurch, Gloucestershire: Dowty Railway Preservation Society
Ashford, Kent: South Eastern Steam Centre.
Beamish, Co. Durham: North of England Open Air Museum.
Belfast, N. Ireland: Belfast Transport Museum.

*Mentioned in this book

Bettws-y-Coed, Gwynedd: Conwy Valley Railway Museum.
Bewdley, Worcestershire: Severn Valley Railway.
Bleadon, Somerset: Somerset Railway Museum.
Blythe Bridge, Staffordshire: Foxfield Light Railway.
Boat of Garten, Inverness-shire: Strathspey Railway.
Bodmin, Cornwall: Great Western Society Ltd. (South West Group).
Bridgnorth, Salop: Severn Valley Railway.
Brownhills, Staffordshire: Chasewater Light Railway.
Buckfastleigh, Devon: Dart Valley Railway.
Bury, Lancashire: Bury Transport Museum.
Butterley, Derbyshire: Derby Corporation (Midland Railway Project).
Carnforth, Cumbria: Lakeside Railway Estates Co.
Chappel, Essex: Stour Valley Railway Preservation Society.
Cranmore, Somerset: Cranmore Railway Co. Ltd.
Didcot, Oxfordshire: Great Western Society Ltd.
Dinting, Derbyshire: Dinting Railway Centre.
Embsay, North Yorkshire: Yorkshire Dales Railway Co. Ltd.
Falkirk, Stirlingshire: Scottish Railway Preservation Society.
Goathland/Grosmont, North Yorkshire: North Yorkshire Moors Railway.
Haven St, Isle of Wight: Wight Locomotive Society.
Haverthwaite, Cumbria: Lakeside & Haverthwaite Railway.
Haworth, West Yorkshire: Worth Valley Railway.
Hereford, Hertfordshire: Steam in Hereford Ltd.
Humberston, S Humberside: Lincolnshire Coast Light Railway.
Llanfair Caereinion, Powys: Welshpool & Llanfair Light Railway.
Lochty, Fife: Lochty Private Railway.
Loughborough, Leicestershire: Main Line Steam Trust.
Lytham, Lancashire: Lytham Motive Power Museum.
Paignton, Devon: Torbay Steam Railway.
Parkend, Gloucestershire: Dean Forest Rly Preservation Society.
Penrhyn Castle, Gwynedd: National Trust, Penrhyn Castle Museum.
Porthmadog, Gwynedd: Festiniog Railway.
Quainton Road, Buckinghamshire: Quainton Railway Society Limited.
Rolvenden, Kent: Kent & East Sussex Railway.
Sheffield Park, East Sussex: Bluebell Railway.
Shepperton, Surrey: Ian Allan Ltd.
Sheringham, Norfolk: North Norfolk Railway.
Shugborough Hall, Staffs (Great Haywood): Staffordshire CC Industrial museum.
Southport, Merseyside: Southport Locomotive & Transport Museum.
Syon Park, London: London Transport Museum.
The Science Museum, London.
Taunton, Somerset: Great Western Society Ltd. (Taunton Group).
Tyseley, Birmingham West Midlands: Birmingham Railway Museum.
Twyn, Gwynedd: Talyllyn Railway.
Wansford, Cambridgeshire: Peterborough Railway Society.
Whitehead, Co. Antrim (Northern Ireland): Railway Preservation Society of Ireland.
York, North Yorkshire: National Railway Museum.